The Woman Doctor's Diet For Women

BARBARA EDELSTEIN, M.D.

The Woman Doctor's Diet For Women

Balanced Deficit Dieting and the Brand New Re-Start Diet

Prentice-Hall, Inc., Englewood Cliffs, New Jersey

The Woman Doctor's Diet for Women
Balanced Deficit Dieting and the Brand New Re-Start Diet
by Barbara Edelstein, M.D.

Printed in the United States of America
Prentice-Hall International, Inc., London/Prentice-Hall of Australia, Pty. Ltd., Sydney/Prentice-Hall of Canada, Ltd., Toronto/Prentice-Hall of India Private Ltd., New Delhi/Prentice-Hall of Japan, Inc., Tokyo/Prentice-Hall of Southeast Asia Pte. Ltd., Singapore/Whitehall Books Limited, Wellington, New Zealand
10

Library of Congress Cataloging in Publication Data
Edelstein, Barbara.
    The woman doctor's diet for women.
    Includes index.
    1. Reducing diets.   2. Women—Physiology.   I. Title.
RM222.2.E27   613.2'5   77-2804
ISBN 0-13-961623-3

For Dr. Stanley W. Edelstein

My husband—that special rare male with enough ego strength to marry a female doctor before it was fashionable and with enough maturity to encourage and enjoy her successes. . . . With love.

Acknowledgments

Carol Saltus, for her thoughtful and intelligent help in editing.
Robert Stewart of Prentice-Hall, for listening to my philosophy.
Joe Steinberg, my attorney in this venture.
Vicky Garcia, my first typist and consultant.
Lorraine Raschke, my second typist.
Judy Johnson, my third typist, organizer, and coordinator.
Diane Schulder, for getting it all started.
David Edelstein, my son, for advice, understanding, and encouragement.

Contents

The Woman Doctor's Diet For Women

Introduction

As a woman doctor specializing in diet, I have been treating overweight people for twelve years. As a woman who has had a weight problem herself, I have been concerned with the subject of overweight for thirty years. Both my personal experience and my practice have kept me aware of how frustrating as well as challenging the problem of obesity can be.

Almost all of my patients are women. I used to think the reason for this was that women prefer a woman doctor. Since then, I have looked into the practices of many other diet doctors, from the simplest to the most sophisticated, and have learned that their patients, too, are almost entirely female. Fat men have it easy! Most of them can lose weight just by giving up dessert. Others warned by their internists to "cut down or else," can effortlessly shed 20 pounds in a month. The rare men who find their way to my office are always the easiest cases I have—I simply put them on a low carbohydrate-high protein diet, and the weight just falls off them.

1

Alas, none of this is true for women. A woman with a weight problem who sets out to conquer it is embarking on a long, tedious, often frustrating, always complicated, and sometimes baffling quest, whose end is always uncertain. Books and articles on dieting flood the market; diet clubs, workshops, and associations flourish. Diet clinics, legitimate and illegitimate, proliferate in all our cities. Yet my office is filled daily with desperate women begging for help. What is the trouble?

I've tried to analyze the problem, and have come up with the following facts:

1. *All important diet books have been written by men.*
2. *Most diet doctors–at least 95 percent–are men.*
3. *Most of the big diet businesses, even if women are involved, are headed by men.*
4. *Most women receive medical assistance, when they need it, from male doctors.*
5. *Most diet books offer only one way to lose weight—either a low fat-high protein, or a low carbohydrate-high fat-high protein diet, or variations on the theme—with exactly the same regime prescribed for both men and women. Or else they impartially present all the diets that have been popular in the past, giving the already confused overweight woman too many choices among diets that may not work for her anyway, because they've all been devised by men using men's bodies as the norm.*

But my twelve years of practice have taught me one major lesson: Women are different! *Especially* with regard to diet, exercise, obesity, weight loss, energy expenditure, and attitudes toward food.

Male physicians have always treated weight problems in women as if they were identical to those of men. Diabetes, pneumonia, and broken bones can be treated this way, but the problem of weight control is much more complex and difficult in women, and requires an entirely different approach which will take into account all the genetic, hormonal, biological, and social differences in the ways men and women handle fat.

Even the effect being fat has on one's life differs for men and women. A fat man is a man who happens to be fat; his bulk is incidental to his maleness and his general capacity to function. A heavy woman is almost a freak—just try getting a decent job, if you're an overweight woman. It's twice as difficult. Or try getting an exciting date, or buying yourself a good-looking dress. If you're a fat woman, whether you're in the ivy halls or the business world, everywhere you meet closed doors, rejection, and innuendos about your size. And when at last you can't bear it any longer and decide to change the way you look, you're constantly hindered and harassed by just those people who should be on your side—friends, family, lovers, husbands—and most discouraging of all, your own doctor.

Here is the very same man who delivered your babies, who saved your life when you developed a blood clot, who stayed up all night with you when you had a raging fever, whom you love and admire, and who may even be a treasured family friend, turning into a hostile, authoritarian bully when it's a question of your weight problem. He utterly misunderstands the underlying physical and psychological causes, and when you fail to lose weight on the regime he sets for you, he accuses you of eating too much, exercising too little, having no will power, lacking character, or even cheating, lying, and being a pig! After every visit you leave his office feeling depressed, inferior, and totally worthless as a human being. I know the problem all too well—I've suffered through it myself, and I hear it a dozen times a day in my own office.

My male colleagues, intelligent, compassionate, and skillful as they may be in every other department of medicine, when they confront the problems of overweight in women become almost uniformly unsympathetic, nonsupportive and stubbornly uncomprehending. They set ruthlessly unrealistic goals of permanent slimness in the face of pregnancy, childrearing, surgery, menopause, aging, and all the other psychophysical crises that beset women at various stages of their lives. "Eat, but stay thin," is the mixed message—or perhaps, "Go ahead and starve, I don't feel a thing." Any way you look at it, the name of the game is sabotage.

How do male doctors sabotage dieting women?

First, by demanding and expecting a permanent, complete cure for overweight. Ridiculous! They are happy if they can keep diabetes and gout under control; why should they expect to cure overweight? In our present state of knowledge, it is a disease that can at best only be controlled. But even to do that much we must be willing to mobilize all the weapons of medicine and psychology available to fight the battle of the bulge; and far too many doctors, instead of using the resources at their command, seem to prefer playing games with their female patients, simultaneously forcing them to meet unrealistic expectations and setting absurd and arbitrary limitations on them.

Take, for example, the doctor who casually gives his patient a

month's supply of diet pills and says to her, "Go forth and lose weight."
When she comes back a month later with a disappointingly small
weight loss, or none at all, he tells her, "Okay, if you can't do better than
that, that's all the help you're going to get from me." Still more common
is the male physician who renounces his responsibility altogether and
sends his overweight patients to diet clubs, conveniently removing the
burden from his shoulders while still retaining a patient.

The other most prevalent form of male medical sabotage is to
judge dieting females by the standards set by dieting males. Women,
being naturally fatter and more resistant to weight loss, of course always
come off a poor second, reinforcing the doctor in his opinion that,
"You're just not really *trying*." Don't be intimidated any longer by
medical male chauvinism. As a woman you think, feel, relate to people,
and move your body differently from men; why shouldn't your pattern
of weight gain and weight loss also be different?

This book explains the difference. It exposes the fallacy of
male-centered dietary theory as applied to women, and shows you
why you have failed so often in the past to get the weight off and keep
it off. By taking advantage of the uniqueness of female physiology—
instead of attempting to deny it, as male doctors seem to do—I can show
you how you can be a successful dieter at any stage in your life: as a
teenager, after you've had a baby, once you've gone on the birth control
pill, after menopause. I will discuss the effects of dieting on all your
systems—hair, skin, intestinal tract, hormones. I will talk about thyroid:
why so many women are anxious to find a thyroid deficiency at the root
of their weight problem, and whether you really need those thyroid pills
you're taking (no matter what you thought, the answer will surprise
you)!

I'll also discuss diet pills—rationally, for a change, without
giving way to the hysteria and sensationalistic scare tactics resorted to
by the male medical establishment in order to block serious discussion
of this complicated issue. You'll see why diet pills were used in the past,
why they're being used now, and what (if any) are the dangers.

I'll talk about exercise, too—why it's so hard to get women to
move their bodies, and why the weight doesn't just fall off them when
they decide to pick up a tennis racquet or go to Slimnastics classes.

And what about fatigue, both before and during a diet—where
does it come from? Why does it fluctuate so? Is it always the result of
being too fat?

I also want to take a good, hard look at fluid retention in

females—is it myth or just a mystery, and why? And menstruation —what does it have to do with your cycles of gaining and losing weight? (It's one of the most irritating and least understood of all the factors influencing a woman's metabolism.) And pregnancy—at no other time in her life do male physicians so infantilize a woman as during these nine months, which may prove critical in establishing disastrous new patterns of uncontrolled weight gain.

You will see that if it's going to work, a diet must be designed to fit your life-style, your likes and dislikes, and your own metabolic and biological cycles. You'll learn when are the times in your life when weight loss is quick and easy, and when it will seem unbearably slow—and how to use this knowledge to become a successful dieter at any time. I'll help you avoid the unhappiness and frustration of dieting strenuously with nothing to show for it—you'll understand why it hasn't been working for you, and what to do. There is a correct formula for every woman to lose weight by, but it may be different for each one; I'll help you find the formula that works for you.

All too often a woman who goes to a male diet doctor for help in losing weight will be put on the diet most in favor with the medical establishment at that moment, and kept on it whether she loses weight or not. If she doesn't, or if her rate of loss fluctuates, of course it must be her fault. Few male doctors are willing to respond and adapt to the rhythms of female physiology. The key to successful dieting for women is flexibility: adjusting the diet to meet the changing needs of a woman's body as it grows, moves through its reproductive cycles, recovers from illness, undergoes stress. This is why I use not one but three basic diets: the Balanced Deficit Diet, the High Calorie Weight Loss Diet, and the Brand New Re-Start Diet.

The Balanced Deficit Diet is my basic diet. It gives you all the nutritional elements you need—fat, carbohydrate, protein—balanced so as to maintain a steady, consistent rate of weight loss. I usually don't believe in playing around with extremes of metabolic imbalance, except that I will occasionally use a short, intense crash diet for fun and variety (see my "Sunshine Diet," Chapter 13, for example). The reason the Balanced Deficit Diet is so successful is that it was especially devised to adapt to the continually changing demands of a woman's physiology. You can adjust it to suit your shifting caloric requirements and still maintain a nutritional equilibrium while you continue to lose weight at a satisfying rate. It's my best basic weight-loss diet for women—I call it my "Core Diet."

The High Calorie Weight Loss Diet is designed for women who need to lose weight even though they are going through specialized and critical situations like pregnancy, puberty, or high physical stress. During these periods women can eat *more* of the correct food, yet still lose weight safely and successfully.

The Brand New Re-Start Diet is my own innovation. It gives new hope to women who have tried everything and still can't lose weight, no matter what they do. *Everybody* can lose weight on the Re-Start Diet! It can even be more effective than fasting. I use it primarily with the dieter who was losing steadily and well, then fell off the wagon at Christmas or while on vacation, and gained a few pounds back. I also use it with the chronic or "yo-yo" dieter who can't get going again. The Re-Start Diet gives these women that extra surge they need to get back on the dieting track, and since it provides all the nutrients the body can possibly need, it's quite safe. Many patients report that within the first couple of days on the Re-Start Diet they experience an emotional and psychological lift, almost a euphoria. They know something great is happening to their bodies and many say they can almost feel the pounds melting off. This extra push launches them into a new weight-loss cycle that usually more than compensates for the weight they gained back.

After you've read this book, I think you'll come away with a new understanding of yourself and your body. You'll realize that your weight problem is uniquely personal because you are a female, because you were born with the potential to be fatter, and because you are YOU! Those are the beliefs you must live by, and those are the beliefs you will lose by!

1.

Men and Women Are Different— Damn the Difference!

As a woman doctor treating dieters, I have discovered that there are three categories of people: overweight women, thin women, and men.

An overweight woman is one who is more than 15 percent heavier than her ideal weight, an excess which usually represents an increase in fat, not muscle. The problem can range from a slight tendency to put on a few extra pounds to a gross obesity that has gone out of control; but once a woman manifests the trait to any extent, she must henceforth always consider herself as having at least the potential to be severely overweight.

Thin women, on the other hand, have been endowed by nature with bodies which burn fat rapidly and efficiently; if they happen to acquire a few extra pounds, they tend to return to normal again very quickly. These women gain weight very seldom, and lose it effortlessly. Listening to them, however, one would think they were constantly wrestling with the specter of obesity. They complain endlessly about the agonies of losing five pounds, while boasting of their expertise as dieters. They can always tell you which are the best, the quickest, the healthiest, the most strenuous diets. Their smug authority would be comical if it weren't for its devastating effect on the fat women who listen to them and feel frustrated and guilty because they can't go home and duplicate their success.

Nothing is more annoying for a woman with a weight problem than listening to a fashionably skinny friend "ex-pound" her favorite diet as she picks up a piece of chocolate cake; it's simultaneously infuriating and depressing. What we need constantly to keep in mind is that it is not superior willpower or self-discipline that keeps such a person thin, but simply the luck of the draw when metabolisms were being passed out.

Then there are men—*all* men. Why don't I divide them into fat and thin, as I did with women? Because from my viewpoint as a diet doctor, all men are the same. A fat man will lose weight as fast as a thin man, once he starts to diet. Almost any man can lose all the weight he needs to by giving up desserts, cutting down on bread and potatoes, and not eating between meals. With these minor concessions, he can lose rapidly, however fat he may be to begin with. Nor is he necessarily expending a lot of energy; he may have a sedentary job and be much less active than a housebound young mother with a weight problem.

Furthermore, for a man to be fat is not the disaster that it is for a woman. No one condemns him as harshly for being overweight, and it has comparatively little effect on his exercise pattern, his work pattern, or his sex life. In all these areas, it's his performance that counts. Many people even think of a fat man as having authority and substance, while a fat woman is considered to be an undisciplined slob.

A fat man is simply a thin man who happens to eat too much. As soon as he stops being a glutton, he becomes thin very quickly. But he, too, would have us believe that his easy and painless weight loss is actually due to the unceasing exercise of willpower. *Willpower,* in fact, is a word devised by thin male doctors, who themselves never have to use it in relation to food but who make sure their overweight female patients hear it constantly, so that they will feel like weak-minded failures. Men and thin women always think they have an abundant supply of willpower, whereas the fact is that overweight women (both currently and *formerly*) have far more of it than the other two categories. However, currently overweight women have more trouble sustaining it for long periods of time.

But the thin male world has conspired to psych women into

believing that they have no willpower at all, which is not only untrue but adds unnecessary guilt to their troubles (serious enough without the extra burden of self-contempt). I've seen so many patients, both male and female, that I can be almost certain that when a fat man walks into my office, he'll be as thin as he wants to be within a few months, without strain or pain, unless he is the rare case with metabolic or deep-seated psychological problems. This is simply not true of overweight women. Uncomplicated male weight loss, like male orgasm, is ridiculously easy and will not be discussed in this book, except as a basis of contrast.

Comparing male and female weight loss is like comparing apples and oranges; they are both fruit, but there the similarity ends. If you're an overweight woman who has struggled unsuccessfuly with diet books and diet plans devised by male diet doctors, I'm willing to bet that a big part of the reason you're still fat is that you've never been told this simple fact: MEN LOSE WEIGHT ALMOST TWICE AS FAST AS WOMEN DO. They burn calories twice as fast for the same amount of exertion. The reason is that a woman's body is naturally composed of a higher proportion of fat to muscle tissue than a man's, and muscle mass burns five more calories per pound to maintain itself than fat or connective tissue.

This means that while a woman's appetite is the same as a man's, she needs only half the amount of food to maintain her weight. How can an overweight woman fail to get discouraged when doctors ignore such a fundamental truth of biology and instead place the blame for her inability to lose weight as fast and consistently as her husband can on her self-indulgence and lack of willpower?

The speed with which men can lose weight is devastating to women. A woman came into my office recently almost in tears over her failure to diet successfully. "I just can't discipline myself—and what makes it all the worse is that my husband can do anything he sets his mind to. He's incredible!" When I asked her, "Do you have something specific in mind?" she replied, "Oh, you know—just whatever he decides to do."

"Do you mean dieting?"

"Well, yes, among other things."

"You mean he's good at losing weight?"

"Yes."

"On the diet I gave you?"

"Oh, no. He couldn't stand that diet."

"Could you stand *his* diet?"

"Well, all he did was cut out bread."

"You call that incredible?"

My point was that when a man diets, he doesn't deserve the gold medal most wives would pin on him. Conversely, when a woman does *not* succeed at dieting, she is not the failure that she and others believe she is.

It is *much, much harder for an overweight woman to lose weight than it is for a naturally thin woman, or for any man.* I used to ask women to bring their husbands with them to my office so that I could enlist their aid and support. I would spend hours talking to them about their wives' physiology and problems with fat storage, only to have them nod in a bored way as they waited for an opening to tell me how *they* lost weight by simply switching from beer to Bourbon or from steak to fish, or when they had to work overtime for a month. It soon became apparent to me that even though these men lived with and loved these women, they could not comprehend what was going on in their wives' bodies. Whether the husbands were of normal weight or overweight, they understood fat only in terms of their own metabolic systems.

If an overweight man cannot relate to the problems of female fat, how can a thin man possibly relate to them? But there he is, his muscles burning up calories, his androgens (male hormones) breaking down fat instead of building it up as female hormones do, his skin tight and elastic, telling women how they should be losing weight (in other words, as quickly and easily as he does.) The thin male doctor is the greatest offender. He's able when it comes to dealing with congestive heart failure, hernias, and pneumonia, but he really isn't interested in helping overweight women; he treats the whole subject as if it were beneath his dignity.

The first clue to his indifference is his recourse to the classic phrase, "All you have to do to lose weight is get some exercise —pushing your chair away from the table!" He refuses to recognize the fact that women are the fatter sex. By the time we're adults, we have twice as much fat as men. We burn from 10 to 15 calories per pound of body weight where men burn from 17 to 20 doing the same thing, yet our appetites are identical. We are classified as overweight if we are between 10 and 15 percent above ideal weight; but ideal weights are usually calculated from life insurance charts, which up to this decade have been overwhelmingly male and oriented toward the upper middle class.

I compute female ideal weight differently. I use 5 feet as a baseline of 100 pounds; for every inch of height over 5 feet, a female adds five pounds. A woman 5'3" tall, for example, should weight 115 pounds. I also take body build into consideration. There should be a difference of about 10 pounds between women with small and medium frames. (Since I see very few Amazons around, I rarely classify females as large.) I also add a pound for every five years past 25 years old. If you are a female 5'5" tall with a small frame, 35 years old, you should weigh between 127 and 132 pounds. You will then be *cosmetically thin*. If you are 15 percent above this weight, you are overweight, *cosmetically overweight*.

The point at which the problem ceases to be one of cosmetic overweight and becomes a matter of health has yet to be determined; that is why I don't impose my own standards of weight on anyone else. Some women want to be superthin, others want only to go down a few dress sizes, still others simply want to feel better. I let my patients decide for themselves what kind of bodies they want; all I am is the tool to help them achieve their aim.

One night I was verbally attacked at a party by a hostile (thin) woman who asked, "Why do you treat people who have only 10 to 20 pounds to lose?"

"In the first place," I replied, "it's their decision, not mine. My patients come to see me, I don't pull them in off the street. Secondly, if women would *stay* only 10 to 20 pounds overweight, they would have no problem at all, and you're right—they wouldn't need me. However, it is the nature of obesity to creep upward. I've seen this happen again and again. Today's teenager weighing 150 pounds will be tomorrow's young adult, 200 pounds. You don't believe that? I've been in practice long enough to see many patients I had as teenagers return to me as young adults, and they invariably come back 20 to 30 pounds heavier because they didn't believe it either. In my office last week, I saw four such cases. Besides, 10 or 20 pounds is a lot easier to lose than 30 or 40 pounds."

Overweight ought to be a disease of prevention. But if it can't be prevented, it should be put under control as early as possible.

Let's look at some of the reasons why women are the fatter sex:

1. BIOLOGICAL—There is no escaping the fact that we were designed as baby receptacles, so nature has seen to it that we will never be without fat. She has decreed that we will always be padded with a soft cushion of subcutaneous (under the skin) fat, in case the fetus needs extra food, protection, and heat. It doesn't matter if you never exercise

your biological function and go through life without bearing a child; nature will pad you anyway, just in case. That is why it's so difficult to lose those last few pounds remaining between you and your lean body mass.

2. HORMONAL—The female hormones that give you your beautiful skin and good bones and protect you against heart attacks are the same ones that make it easier for you than it is for a man to convert food into fat. Estrogens and progesterones are naturally fat-producing and fat-hoarding hormones. Even if you have your ovaries (which produce the hormones) removed, the adrenal glands will take over and secrete estrogenlike hormones for the maintenance of body fat.

If you're taking birth-control pills (whose operative ingredient is estrogen), you will be 10 percent more likely to convert food into fat if your weight is normal, and close to 20 percent more likely if you tend toward overweight—and this is not even taking into account the fluid-retaining properties of both estrogen and progesterone. These propensities for making you fatter are present in all female hormones, both natural and synthetic.

3. SOCIAL—*The "Home-Baby-Eat Syndrome"*—The responsibility of feeding a family three meals a day means having to think about food morning, noon, and night—buying it, preparing it, seeing that it gets eaten, cleaning up afterwards. Every housewife spends a lot of time in the kitchen, where she is constantly exposed to temptation, especially if she prides herself on her cooking. It's easy to deceive oneself into thinking that it's all for the sake of one's husband and kids, but how many good cooks don't also love to eat? Still, you can't put all the blame on proximity. A chronic overeater will manage to get her excess calories somehow, even if she works outside her home during the day; she'll overeat when she gets home at night, or she may spend the weekend baking bread and cookies to make up for lost time. Maybe her husband enjoys cooking as a hobby; then of course she has to eat everything he makes, so as not to hurt his feelings. It will be interesting to see what happens now that sex roles are being reversed in so many areas. If men move into the kitchen, will we have thinner females and fatter males? And what will happen when the formerly thin grandmother is cast in the role of baby-sitter for her working daughter? Will she also become overweight?

I feel, though, that the social role of wife and homemaker is not as important a factor in obesity as women would have us believe. Inappropriate responses to food cues plague the would-be dieter, whether she is in her own home, or another's, or at a job. Since food is always available—if only from the candy machine at the office—the overweight woman will always find a reason why she should eat it!

4. BODY MAKEUP—As if all the above didn't give men a sufficiently unfair advantage, women also require fewer calories. Some authorities claim that women require 2 calories less per pound of body weight than men, but I think it is actually closer to 5 calories. The reason is that more calories are needed to sustain large muscle mass in a male than to sustain fat in a female. Men are usually heavier and taller than women, but even the smallest man has more muscle per unit of weight than the largest woman.

5. APPETITE—Nor has nature even bothered to equalize the difference in the way men and women burn calories by giving the woman a smaller appetite. Appetite, unfortunately, depends entirely on the individual; so many psychological variables influence hunger that it is almost impossible to measure appetite objectively. All we can be sure of is that a woman can, and will, often eat as much as or more than a man, even though she requires fewer calories.

Everything I have been saying here applies to all women. Multiply it by two, add a triggering mechanism for overeating sugars and starches, stir in a dash of carbohydrate intolerance, and you have the stew in which the overweight woman finds herself.

2.

The Three P's—Puberty, Pregnancy, Pill—and Other Problems in Fat Formation

According to current thinking , there are two ways people can become fat. The first is by *increasing the number* of fat cells present in the body, the excess cells being acquired either *in utero,* during infancy, or in puberty. This is probably the most common way of becoming moderately to severely fat, and is called hyperplastic obesity; most people who are severely fat (over 30 pounds) have too many fat cells.

The other way to become obese is by *enlarging the existing* fat cells already present in the body. This is called hypertrophic obesity, and probably accounts for small weight gains of up to 20 or 30 pounds.

In one experiment, thin male volunteers were fed 8,000 calories per day of a high-carbohydrate diet, and gained about 30 pounds in six months. When the increase in fat was measured in these volunteers, it was found that the size of the fat cells had increased, but the number had remained constant. When the excessive feeding stopped, the men easily dropped back to their normal weight by reversing the process, decreasing the size, not the number, of fat cells.

15

But why did the men gain only 30 pounds? If it takes 3,500 calories to make one pound and they were in excess by at least 4,500 a day, they should have gained 1½ pounds per day, 10 pounds per week, 30 pounds per month, and 240 pounds in six months. This leads me to question the validity of calorie counting alone as an index of weight gain, and also to wonder if there are perhaps certain people who will never gain weight, no matter how much they eat. One is forced to conclude that a person who is not genetically programmed to be fat—that is, who does not contain an excess number of fat cells—will never become severely fat, no matter how many calories he or she consumes.

This study intensifies my current feeling that the tendency to be very fat is probably inherited. However, this ability may or may not be manifested, depending on the environment in which one is raised. A child with the genetic potential to become fat, born into a family that forces her to be athletic, will not manifest her fat trait until and unless she becomes more sedentary. The same child born into an inactive, slow-moving family in which eating is an important ritual will manifest a weight problem much earlier in life.

The younger the age at which the child becomes fat, the more difficult the problem of controlling her obesity will become. Once the potential to be fat is established, there is no limit to how fat you can get.

Several questions arise at this point: When are the fat cells formed? If they are laid down *in utero,* then do we totally defeat our purpose by creating babies who are fat at birth? (This will be discussed further in the chapter on pregnancy.) If, on the other hand, fat is formed in the first two years of life, should we not start babies on skim milk or unsweetened fruit? How do we know which babies will be fat or thin? We can never be sure; we do know that if one parent is fat, there is a 40 percent chance that the offspring will be fat. If both parents are fat, there is an 80 percent chance that the offspring will be fat. We also know that most females have far more fat potential than most males; therefore, daughters in families in which either parent is fat should be encouraged to watch their food intake at all times.

Puberty (which I set at ages 8 through 14) in a female is another crucial period in the creation of new fat cells. A child who is overweight at puberty has the potential to be fat for the rest of her life. Unfortunately, girls usually don't undergo the rapid growth spurt that boys do at this time. Their growth spurt is shortened by their newly produced estrogen, which also enhances fat formation. Sometimes a heavy child will grow very tall and become very slim when she reaches adolescence. But her growth stops before she stops laying down new fat, so at the age of 15 she will suddenly start to become heavy again. At this time the natural tendency for fat to be laid down in the breasts and hips is exaggerated by overeating. Figure problems that start during puberty remain with the female her entire life.

There is no such thing as "baby fat" in a female, but it's a bit of wishful thinking that dies hard. I still have mothers come in with 13-year-old daughters weighing 150 pounds, waiting for the baby fat to disappear. Puberty is a dangerous period for girls with overweight potential. But since puberty is also a time of growth, you should try to achieve at least weight maintenance during these years. This is probably the best you will be able to do, because the body protects fat during this period of growth, making weight loss extremely difficult.

Puberty and adolescence are the times when mothers run to the pediatrician hoping to be told that their overweight daughters have a glandular problem. They almost never do: in my experience, 99 percent of adolescent obesity is based on a combination of genetics and overeating. But to be absolutely safe I always order a thyroid test. Not only does it reassure me, it makes the parents feel better.

I don't work with adolescents in a formal program because they are usually not strongly motivated enough to follow a definite diet. If a woman is seeing me for a diet and expresses concern over her young (8 to 13) overweight child, I ask her to bring the child along. I invite the child to listen to my discussion with her mother, and encourage her to ask any questions she may have of her own. I also offer to weigh her. At the end of the mother's visit, I tell the child that she is welcome to come to the office at any time, to get weighed or merely to listen and ask questions. Children usually respond very well to an informal program of this sort, since it is not threatening and no expectations are placed on their performance. Many of them maintain their weight level very gratifyingly and in the process also learn more about food and its action in the body.

If girls were like boys (but they're not, and we can't ever afford to forget it) the major fat-depositing phase of their life would be over after adolescence. But girls have to deal with two more major fat-depositing phases.

Pregnancy or the Pill: What Your Gynecologist Won't Tell You

How many times have I heard "But I *was* thin until I got pregnant!" or "My doctor told me I would put on only a few pounds with the pill, and I gained 15 pounds!" When the specific problems of obese women come to be studied seriously, pregnancy will probably emerge as the number one villain. During those nine months it seems as if the whole body has gone into overtime producing fat, as nature pulls out all the stops. I will discuss dieting and pregnancy in another section of this book. Here I'll simply say that in my opinion, pregnancy surpasses even puberty as the most critical period in the female fat cycle.

A newcomer to the legions of the fat woman's enemies is the birth-control pill. I only wish women could keep young, sexy, and infertile with the male hormone testosterone, which doesn't influence fat formation. Many thin women respond to the birth-control pill only by picking up a little water weight, easily got rid of with a diuretic. In the fat-prone female, however, the birth-control pill, at least initially, acts like a mini-pregnancy, causing the body to produce increased amounts of *fat* as well as water. Therefore, if you are going to start taking birth-control pills, it is imperative that you also start on a diet simultaneously and try to decrease your daily caloric intake by at least 10 percent. Remember, too, that a birth-control pill by any other name acts in the same way; anything containing estrogen, whether it's used after surgery, or for your skin, for cramps, or to regulate your periods, will cause the increased deposition of fat.

The three P's—puberty, pregnancy, pill—are the most important factors in female fat formation, but there are other significant times when a female predictably will put on weight. Being aware of them will help to circumvent the danger.

Leaving Home

A girl is very apt to put on weight when she leaves home and goes away to college or to work. Colleges are the worst offenders; the old dormitory method of feeding all the girls the same set meal is murder for anyone with overweight tendencies. How well I remember my own college days. My mother, herself still a size 12, struggled to keep me thin as I was growing up. Somehow she managed to hold me at 15 pounds overweight through high school by insisting that I eat mainly broiled meats and vegetables. I had never seen a casserole until I walked into the college dining room, and then I went wild! While everybody else complained loudly about the food, I could hardly wait to gorge myself on buttery biscuits wrapped around meat and smothered in a cream gravy. Food is generally much better in colleges these days; many schools provide salad bars and skim milk. But for economic reasons they still fall short on low-calorie sources of protein. All colleges should go on a cafeteria system that would offer a choice of plain hamburgers, plain broiled chicken, and plain fish, along with high-caloric foods. Or overweight students should try to find accommodations where they have some control over the preparation of their food.

Gaining Weight After Surgery
Another significant period of weight gain in a woman's life is after surgery, such as hysterectomy, gall bladder, or hernia repair. Weight gain is usually due to a decrease in activity plus an increase of overeating, together with a slowing down of metabolism. The latter probably represents a period of hormonal exhaustion following the stress of surgery. When my patients tell me they are going into the hospital for surgery, I give them the following instructions:

1. Ask your doctor to give you an 800-calorie, high-protein diet in the hospital after surgery.

2. While you are in the hospital, if you must be given I.V. (intravenous) fluids, *beg* to have them with the least dextrose content possible.

3. Find out what is the earliest possible time you can get moving, and don't let that male doctor infantilize you—the day of the six-week postoperative convalescence is over!

4. When you get back home, eat as much protein and as little of anything else as possible.

5. Take lots of vitamins, especially B-complex and C.

6. Use the postoperative period to make your body more beautiful, not fatter. Just imagine you are taking a vacation, and spend it in the bedroom and the bathroom, not in the kitchen. Emerge from your convalescence looking as if you've just come back from a beauty resort.

Menopause
Menopause is a more serious problem for thin women, since fat women have usually gained as much weight as they're going to by this time. The mechanism of weight gain in menopause appears to be the enlargement of existing fat cells. Decreasing calories 5 percent below the number needed for maintenance for every five years after 40 should keep weight within normal limits.

3.
The Best (and the Most Difficult) Times of Life for Dieting

1. MIDDLE TEENS (HIGH SCHOOL)

This is an excellent time to attempt a *serious* diet. Growth in the female is usually complete. Menstrual irregularities have usually straightened out, and fluid retention is not yet a problem. Motivation is often strong, especially if there is an interest in boys. Stubbornness, so characteristic of adolescence, can be harnessed to the service of vanity and become a useful adjunct in diet therapy. Most important of all, fixed patterns of eating have not yet taken hold and the "forbidden food" syndrome has not fully developed.

This is one of the few fixed neuroses that accompany chronic overweight. It can best be illustrated by a luncheon date I once had with a very skinny friend. When she ordered a shrimp cocktail and black coffee, I looked at her in amazement. "Why are you dieting?" I asked.

"I'm not dieting," she said with genuine amazement. "I *like* shrimp cocktail and black coffee."

"But—but," I stammered, "if I looked like you, I would be ordering a hot roast beef sandwich with French fries on the side."

21

"But I don't want them today," she said. "I can have them any time, so I don't need to have them today."

Of course she was right. When you can have as much of any food as you want, it has no meaning to you. Chronically overweight women can *never* eat as much as they want, so high-calorie (forbidden) food is very precious to them. The foods with high carbohydrate levels and pleasant taste which we love as children later become the familiar cravings of adulthood. Such cravings are easier to break before they become firmly established. With my teenage patients, I try to substitute a large quantity of the right food for a small quantity of the wrong food.

Interestingly, many parents worry when their child eats large quantities at one time, no matter what the food is. I've heard many mothers complain "She eats so much meat," or "She eats two bunches of carrots at one time. How can she possibly lose weight?" Mothers have to learn that it is far better for a child to sit down and eat a *large* meal of salad and meat than to eat many small meals of cookies, candy, and soda (which the parent may never see). It's much better to eat large amounts of the right food publicly than to sneak small amounts of the wrong food when nobody's looking.

How Much of What Should Teenagers Eat?

It is hard to keep quantity down in the diet of a teenager, but the parent should wisely and *silently* substitute food that will be satisfying and as low in calories as possible. If the main course is all starch, than a salad should be served first. This helps fill up the teenager before the fattening part of the meal. Soups that are free of fat are also useful to serve a hungry teenager before a meal. In the teen years before the normal appetite-hunger complex is destroyed, a feeling of fullness really registers "stop eating."

Teenagers are also usually very willing to eat whatever is put in front of them unless it is something that they intensely dislike. Many mothers object to the fact that their daughters have limited tastes. However, if they like one good source of protein, like chicken or beef, and one low-calorie vegetable, that is all they need. Too often as parents we want to impose our own feelings and attitudes about food on our children.

Dieting Teens and Their Mothers

Since dieting teenagers are so dependent upon their mothers, we ought to take a look at these influential females. I divide mothers of dieters into two major categories: (1) the uptight, anxious mother, and (2) the saboteur.

The uptight, anxious mother comes into the office hovering nervously and defensively over her daughter; the subject of diet has created a tremendous amount of bad feeling between them. She is usually quite thin and well groomed, in contrast to her daughter, who is sloppy and overweight. The uptight mother sends out two messages: shame at her daughter's appearance and an intense desire to make her lose weight. Unfortunately, what the daughter picks up (and amplifies) is the shame. Every mealtime becomes an argument, if not a pitched battle. The harder the mother tries to make her daughter diet, the more the daughter eats. (This sort of situation lays the ground for the pathology of the passive-aggressive personality, to be discussed later.) In such a case the mother and daughter must first be separated emotionally before anything else can be done, and if necessary I act as referee. I tell the mother that she may come into the room to listen to my discussion with her daughter, but only as an observer; she is not to interrupt or interfere. But I do want her to hear exactly what I say to the daughter, especially concerning what she is to be allowed to eat, so as to avoid arguments between them later.

The mother's role is to supply the food; the daughter's responsibility is to eat what is supplied. Those roles must be faithfully executed, and to make things easier, I try to take the pressure off both mother and daughter. I tell the mother to let up, and I tell the daughter to settle down and diet. I tell the mother that if she's upset at her daughter's eating habits she should talk to her calmly about them, not scold or shout at her. I try to break the vicious spiral between them, in which anger and hostility cause compulsive overeating, which leads to more anger and hostility, and so on ad infinitum.

The second type of mother, the saboteur, is more difficult to control because she is so sweet. She is usually well-meaning, genuinely sympathetic, and honestly does not dislike her daughter for being obese; her problem is that she is annoyed with the world for not appreciating all the good qualities of her child. While the uptight mother insists on bringing her daughter in to see me, the saboteur usually comes to the office only at the insistence of her daughter.

You can always tell a saboteur by the way she says, after you've outlined the rules of the diet, "But you must remember, there are other people in my household I have to feed. I can't cater just to one child." I try to explain that the whole world caters to the thin person:

every restaurant, every short-order coffee shop, every school cafeteria, every ski lodge, every resort hotel. In all these places, as well as the local drugstores and supermarkets, the thin person can always get adequately fed, but nobody caters to the fat person. Sometimes I say, "If your child were an alcoholic, would you place a bottle of Scotch on the table and say, 'I'm sorry, you can't have any, but your brothers and sisters want a drink,' " or I ask, "How can you bring something to the table that is so excruciatingly tempting to an overweight teenager and expect her not to eat it?" Most mothers get the idea.

But the saboteur has many other tactics. I remember a patient of mine, a teenager who failed to lose weight one week. When I inquired why, she looked accusingly at her mother: "She's been putting meat and cheese and mayonnaise in my sandwich at noon." I had told the mother to give her only plain roast beef or chicken sandwiches with either mustard or catsup. When I asked why she had not followed my instructions, she replied, "The meat looked so lonely sitting there all by itself on that poor little piece of bread."

The worst case of sabotage I can remember was perpetrated by Sandy's mother. Sandy was about 5'2" and not terribly attractive, mostly because she weighed 175 pounds. She came from an upper-middle-class family and was highly motivated to lose weight. Sandy dieted in a sensible way for six or seven months, and lost about 65 pounds. By the end, she looked marvelous. About a year later, I saw her again; she'd gained it all back and then some. I was astonished. Her attitude had been good throughout her diet. She never felt sorry for herself; she had gone on maintenance smoothly; she had taken up various sports to keep her new figure. "What happened?" I gasped. Sandy told me the following story:

"When I got down to 108 pounds, my family decided I looked sick. There wasn't a day went by that they didn't say, 'You've got to eat more; you're going to get sick.' " Her mother would constantly bake cookies and cakes for her. Her father would tell her how pale and unattractive she looked (which was untrue). They nagged and harassed her relentlessly, until finally she gave in, deciding it was easier to eat than turn every meal into a power struggle. Within a year, she had eaten herself back to 180 pounds. The saddest thing was that Sandy was just about to go to college and start a new life, and here she was 70 pounds heavier than she had been before she came to see me.

Sandy's story is not as unusual as you might think. It's an extreme case, of course, but it happens in varying degrees to many teenagers who lose weight. The teenage years are a very favorable time for dieting, but good eating habits have to be established by responsible, knowledgeable parents who understand that to be thin is healthy, and who really want their child to be thin for the sake of her own future happiness.

Crash Diets

Teenagers love crash diets, and the crazier the diet, the better they like it. A crash diet is a diet designed for a rapid weight loss in a short period of time. It has nothing to do with sustained weight loss and is useless for older women who lose weight more slowly, or women who have more than 15 pounds to lose. You also learn nothing about nutrition from a crash diet. But a teenager can often use the stimulus of a crash diet to start on a regular diet. For instance, a week of bananas and skim milk, or of steak, eggs, and tomatoes, or of cabbage soup will often appeal to her. However, she must understand that afterwards she will have to proceed to a more sensible kind of dieting if she has more weight to lose.

2. AGES 20 TO 40 (YOUNG ADULT AND ADULT)

The best years for losing weight are between the ages of 20 and 40. Most of the significant changes in your life occur at this time, which is also the time when motivation, determination, and hormones are all working for you. You have the maturity to handle a new body; you know how to make it serve you. You are in a position to decide how you're going to let it change your life. Losing weight as a teenager can be a lot of fun, but not until you've become a young adult can it be a truly gratifying experience.

3. AFTER 40

The forties are troubled years, because for some reason your body doesn't want to obey you anymore. Your skin doesn't fall into place the way it used to after you lost weight; your body changes shape; veins appear in your legs; weight loss starts to slow down. Your tightly knit world begins to shift or even disintegrate—your family structure may alter through death or divorce. More and more situations in life seem to be beyond your control.

 The key to weight loss in the forties is *concentration*. You must have the ability to focus on losing weight and to block out anything that's extraneous to your purpose. You must realize that it is more important than ever for you to take care of your body, and that you still can be a sexual and seductive human being. The forties are an anxious time, but staying slim can only make them easier.

Weight loss in the fifties and sixties is often incredibly slow, and this discourages women from staying on diets. However, it so markedly improves the way you look and feel that it is worth every bit of extra effort. I like to say that for every ten pounds you lose you become a year younger. Exercise is the key to increasing the speed of weight loss during these years—not sporadic spurts, but sustained, brisk walking, or even jogging. Many women of this age prefer swimming, but swimming pools are not always available, and this form of exercise can be exceedingly drying to the skin.

The fifties and sixties are a good time to think seriously about having cosmetic plastic surgery, particularly in the face and neck, after you've achieved a major weight loss.

GOOD AND BAD TIMES TO LOSE WEIGHT ACCORDING TO AGE

TEENS

Favorable
1. No ingrained hunger-appetite response.
2. "Forbidden food" syndrome is not firmly established.
3. Too lazy to fix own food.
4. Suggestible, and does well following instructions.
5. Fluid retention not yet a problem.

Unfavorable
1. Time of growth.
2. Motivation and discipline lacking.
3. Must rely on others for food.
4. Lack of knowledge about nutrition.
5. Peer group pressure.
6. Activity level unpredictable.

TWENTIES AND THIRTIES

Favorable
1. Growth has stopped.
2. Motivation is excellent.

Unfavorable
1. Marriage.
2. Pregnancy.

3. Hormones stable.
4. Activity levels comparable to all other age groups.
5. Other interests have developed.

3. Children.

FORTIES

Favorable
1. Last-ditch try for sexuality.
2. More fat storage around organs—so scale weight isn't an accurate reflection of how you look.
3. Start losing taste for sweets, which put on weight the fastest.

Unfavorable
1. Skin not as resilient.
2. Psychic energy directed toward staying well.
3. Emerging fear of disease.
4. Alteration in state of well-being.
5. Appetite = health.

FIFTIES

Favorable
1. Comfortable age for women to give up trying to be a sex symbol and concentrating on feeling and looking good, and working hard.
2. If she is very healthy, it is a good period.
3. More realistic expectations of change.

Unfavorable
1. Skin tone poor.
2. Weight loss slower.
3. "Grandmother" syndrome.
4. Changes in family structure; stability of core and extended family threatened.

SIXTIES

Favorable
1. No new fat being produced.
2. Normal process of aging made easier.
3. Disease processes respond well to weight loss (arthritis, diabetes).
4. Rebirth of sexuality.

Unfavorable
1. Weight loss becomes secondary to survival (chronic disease).

4.
The Universal Overweight Syndrome

While I was writing this book, a former patient of mine worriedly asked me if I were planning to include case studies of my patients. I smiled to myself as I reassured her that I would not write about her particular case. I did say, though, that she might recognize herself in parts of my book because there is a universal overweight personality syndrome. In fact, overweight women have so many characteristics in common, it's amazing that no one has bothered to document them.

The purpose of this chapter is to lighten the psychiatric case against overweight women. I think the traditional "fear of failure" analytic jargon that locks her into having unconscious reasons for being fat is overdone.

Psychiatrists have been very successful at conning the overweight female; they've managed to make her as well as everyone else believe that her overweight problem is based solely on her libido (sexuality). One very popular theory in psychiatric circles is that the overweight female retreats behind a barrier of fat because she is unable

29

to deal with her sexuality. Another holds that she eats to fill the emotional void within her, using food as a substitute for love. In fact, as both a former psychiatric resident and a general practitioner, I find the overweight female to be relatively free of the neurotic tendencies that I see in women in the general population. Phobias, anxieties, hysterical features, and compulsive behavior are comparatively uncommon in the universal overweight. Their eating behavior, the segment of their personality that is deranged, does not seem to have prevented normal adjustment in other areas of life.

But the universal overweight has certain personality quirks that are important to understand because they keep getting in her way when she tries to lose weight. She exhibits a great many traits of the passive-aggressive personality whose genesis was described in the previous chapter. This personality type opposes change by passive resistance rather than by open confrontation. Such behavior represents, for the overweight woman, a carry-over from the way she met difficulties in childhood, and while in other areas of life she may be, and often is, extremely mature, she reverts to this type of behavior when dieting. She becomes stubborn, she procrastinates, she erects all kinds of obstacles, and ultimately succeeds in failing to lose weight. Many women with this syndrome are completely unaware of what they are doing until it is pointed out to them.

The infantilism of this behavior was made obvious to me by a patient who came to a party at my house. As I usually do, I served a rich dessert. I overheard my patient say to her husband, "If the doctor thinks I'm not going to eat that, she's crazy!" She was rebelling against me as if I were her mother, thirty years later, instead of her doctor. While this woman appeared to accept what I said and seemed to be an ideal patient, in reality she was doing all she could to undermine my efforts to help her become thin.

Another example of the passive-aggressive personality is the patient who came into my office in tears, saying she would do anything to lose weight. I told her that wouldn't be necessary, all she had to do was diet. But when I explained what she would be eating for lunch, she interrupted me. "Wait a minute, doctor, you don't understand. I eat my lunch in a cafeteria." "Haven't you ever heard of a brown paper bag?" I asked. She appeared horrified at the prospect of packing her own lunch, but minutes before she had said she would do anything to lose weight.

I think the ambivalence that the female feels between the desire to be thin and the inability to do so is not so much a fear of being thin, as it is directly related to hunger, temptation, and motivation.

Ambivalence in dieting stems from realistic *present* conflicts, not sexual repression.

More than she realizes, the universal overweight needs to be in control of every situation; she is *particularly* resistant to giving up control of the quantity and kind of food she eats. Consider the "dinner-party dieters," those women who habitually go off their diets at dinner parties because (so they say) they don't want to offend the hostess. I used to tell them, "Go ahead and offend the hostess. It's your body, not hers." But that line of reasoning didn't work. The passive universal overweight is too anxious to please people—at least on the surface. So I found a new angle. I would ask the dieter to name a food she disliked intensely—raw oysters, for instance. Then I would ask her what she would do if her hostess put a plate of raw oysters in front of her. "I wouldn't eat them, I hate oysters," she invariably replied.

"But wouldn't you be afraid of hurting the hostess's feelings?"

"No, that wouldn't bother me. I really hate oysters."

Then I suggested that whenever the hostess offered her food not allowed on her diet, she imagine it was something she disliked. That would permit her to refuse to eat it without fear of offending the hostess.

Universal overweights would like you to believe they are helpless victims of a world intent on stuffing them with food, but in fact, as the above examples show, they are in full control of their food intake.

Another problem with the passive-aggressive personality is an inability to follow instructions. As part of my routine office procedure, I ask my patients to write down their diet themselves as I dictate it to them so that there will be no mistakes about what I've said. I keep my Core Diet relatively simple and easy to follow. I say, *"This* is how I want you to eat, and there are *no* exceptions." When they return for their next visit and I ask them if they have followed the diet to the letter, about 30 percent of the time I get answers like, "Just about," or "Almost." When I ask, "Why 'almost'? Don't you remember I said, no exceptions?" the responses are, "I couldn't help it," or "I lost my list," or "I forgot what you said."

Many patients, even if they start out following instructions, end up after a few days deviating from what I originally told them. They add a drink (or two, or three), or they fill up on fruit between meals. When they come back to the office and show no weight loss, they are surprised. If they could only follow my instructions—*no* additions, *no* substitutions—my patients would all be successful dieters.

There is also a strong tendency on the part of the universal overweight to rationalize—to try to fool both herself and the doctor whenever she can. She finds reasons to eat fattening foods, such as crackers when her stomach is upset, ice cream when she has a sore throat, and sweets when she feels tired. She also finds a new set of reasons not to eat diet foods: "Tomato juice gives me heartburn," "Eggs in the morning make me sick," "Cheese constipates me," "The kids drink all my diet soda."

The universal overweight gets so much pleasure from food that when she is deprived of this pleasure she becomes edgy, irritable, and unhappy; I have come to the conclusion that food is the great tranquilizer for these women. Their concern for food and its taste is always at a high level, and their absolute refusal to eat anything that is not pleasurable to them, even though they know that it will result in a greater and more efficient weight loss, shows how stubborn they can be. I repeatedly tell them that a diet is not exciting, it is boring; that it is meant to be tolerated, not necessarily enjoyed; that it is not going to please their palate; and that they are going to have to find their pleasure in other things besides food. (I do, however, offer some tricks—see Chapter 6—that keep the boredom to a minimum.)

Nevertheless, it has been my experience that the universal overweight is a good deal happier than we would expect her to be. One of my patients put it this way: "I really don't know what I'm doing here—I'm perfectly happy. My relationship with my husband is great, I'm taking a course at school I enjoy, my kids are fantastic. Why do I overeat?" It seems that universal overweights are so happy in general that most of the time they feel it is not worth the trouble to lose weight. They are able to compartmentalize their body image and their weight problem so that they feel it to be a problem only when social, family, or working situations make life uncomfortable for them. Then and only then they become aware of the anguish of being overweight. But as soon as they can encapsulate their weight problem by denial or by manipulating the environment so as to reduce the painful awareness of their obesity, the universal overweight will effectively tuck the problems away and resume being quite happy again.

I have often been asked if I think overweight is a disease like alcoholism. I used to reply, vehemently, "Absolutely not!" Now I am beginning to think that alcoholics and foodaholics do have certain tendencies in common. Just as the alcoholic cannot stop at one drink, the foodaholic cannot stop at one carbohydrate. The alcoholic is psychologically dependent upon alcohol when he is under tension or depressed, and the foodaholic eats for the same reasons. However, the internal needs of the alcoholic are much greater than those of a person who overeats; the level of personality disintegration in obesity rarely, if ever, reaches the level that you find in alcoholism. Mental acuity is never lost as a result of overeating, and mental and physical deterioration is rarely as severe as with alcoholism. In addition, alcohol provides mood alteration, an escape from reality, and a cover for rage, frustration, and sensitivity. Food, on the other hand, does none of these things.

Should an overweight woman go to a psychiatrist? Only if she has other emotional problems, my psychiatrist friends tell me. And they should not go expecting to lose weight, but only to gain a better understanding of their reasons for overeating.

To sum up, the universal overweight is a passive-aggressive personality type who resists losing weight by passive resistance rather than direct confrontation, by rationalization, and by manipulation. It is important for her to understand the nature of this pathology because it blocks her effectiveness as a dieter. This, combined with her intolerance for long-term self-control and her distorted hunger cycle, is the psychosocial and behavioral explanation of why weight loss is such a frustrating problem for her.

5.

Why Does a Woman Start Dieting?

A woman with a chronic overweight problem will start a major diet (losing 20 pounds or more) on the average of once every three years. The decision to take such a drastic step is most commonly triggered off by one or more of the following:

1. *psychic pain*
2. *clothes*
3. *husbands or boyfriends*
4. *physical symptoms*
5. *general health*
6. *job discrimination*
7. *dating*
8. *affairs and affairs*

1. Psychic pain and the pain-pleasure principle

Psychic pain is a state of acute discomfort at being fat. When she reaches this stage the patient hates food, hates people, and hates herself, a situation which is doubly painful because normally she loves them all. If the psychic pain that launched the diet could be remembered throughout it, few dieters would fail. Lately I have been asking my patients to make a tape recording of how they feel before they start a diet. Then I ask them to replay the recording every day until all the weight is lost; otherwise it's too easy for them to forget their reasons for dieting in the first place. To be effective as a motivation for losing weight, the degree of psychic pain at being fat must exceed the pleasure of eating fattening food. It's a seesaw effect, like this:

SUCCESSFUL DIET   *UNSUCCESSFUL DIET*

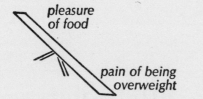

Psychic pain is generated by the unpleasant way you look, the disgust you feel at yourself, and the embarrassment and revulsion you *think* others feel toward you. As with physical pain, people have different levels of tolerance for psychic pain. Some people who have low psychic pain thresholds will let their weight go up only 10 to 15 pounds; others with high psychic pain thresholds can allow their weight to go up 100 or more pounds above normal. In addition to keeping psychic pain levels high during dieting, it might be useful to re-educate those people with high psychic pain thresholds to respond earlier to the need to diet.

2. Clothes
Despite my running feud with designers who refuse to make attractive clothes for fat women (except for maternity dresses), I must admit that clothes are one of the major incentives for women to diet.

"I've just eaten myself straight through my last dress," a patient will tell me, or "I can't afford a whole new wardrobe," or "I'm sick of those old-lady styles which are all they've got in my size," or "I'm going on vacation, and I have *nothing* to wear." (You hear this from thin women, too, but fat women really mean it.) Sometimes a good hard look in a three-way fitting-room mirror sends a woman running to me for help. But it's easy to exaggerate the importance of clothes size alone; scale weight should be the index for deciding whether or not you should lose weight. There are women who can weigh 160 pounds and yet wear only a size 12, but overweight women like to think of themselves in sizes instead of pounds. When I ask my patients what they weighed when they got married, they invariably answer, "I was a size 10," as though that were all that mattered. Nor are clothes always a good indicator of fat loss. Your clothes may feel looser from redistribution when you haven't lost an ounce.

3. Husbands or boyfriends

As a rule, the heavier the woman, the less effective are her husband's or boyfriend's feelings as a motive for staying thin. Many a 200-pound woman has come into my office who, when I ask how her husband feels about her weight, replies, "He doesn't say anything," or "He never mentions it." If a husband is going to object, he will do so loudly and emphatically at the first 20 pounds; after that, he buries his head in the sand. Most of my extremely obese patients report that their husbands are concerned about their health, not their weight. Boyfriends, unfortunately, are a poor influence on overweight women because they have chosen them *in spite of* their obesity or even *because* of it, and are not going to put any pressure on them to diet.

There is, however, a familiar phenomenon known as the "Wedding Gown Dieter," a girl who will try very hard to lose weight so that she can get into a smaller size wedding gown. I can't understand why people who sell wedding gowns don't recognize this fact and make a point of selling a gown two sizes too small to all overweight customers who tell them they are going on a diet. A girl *will always* lose weight for her wedding; all of my patients needed to have their gowns altered one, two, or three sizes smaller by their wedding day. But unfortunately, when the honeymoon is over . . .

4. Physical symptoms

Specific physical symptoms bring many women into my office. Usually they are: shortness of breath (when climbing stairs), fatigue, swelling of ankles, nonspecific bad backs.

Most of the time, these symptoms are due mainly to increased bulk and improve dramatically with a small weight loss of about 10 to 15 pounds. The exception is fatigue, which I will discuss in detail later.

5. General health

Many women are sent to me by their orthopedic surgeons for help with back injuries that won't get better, and I see a number of diabetics whose blood sugar is difficult to control because of their weight. Rarely do I see patients with heart disease; I like to assume that their internists are managing the weight loss in these cases. One would think that women who know that losing weight will dramatically improve their health and lengthen their lives would be the most successful dieters; unfortunately, they have the highest rate of failure. To the universal overweight, food may be more important than life itself.

6. Job Discrimination

Where do you find most overweight women working? Not in the glamour jobs, I assure you! You find them in routine civil service jobs or relegated to the dusty back rooms of drab businesses—never working in films and television, or as airline stewardesses. Yet a prospective stewardess's weight should be a consideration only if it might prevent her from getting through the aisles. I do hope that overweight women will assert themselves in the realm of job discrimination and job opportunity, exposing those employers who are really interested mainly in hiring sex objects. I hire my office aides for their ability alone, not by what they weigh on the scale, and pressure should be put on equal-opportunity employers to do the same.

But for the present, most fat girls can't get the job they want, especially if they are unskilled or have limited skills to bargain with. It's a shame, because overweight girls are usually steady, dependable and sick less often than their thin counterparts; I find they make excellent help.

7. Dating

As you might expect, heavy girls date less. They do have friends, and often a boyfriend; many of them marry childhood sweethearts. But they seldom have much choice.

Heavy women are a drag on the dating market, and they know it. What they don't always realize is the incredible difference losing weight makes in their attractiveness to men. It doesn't matter a bit how pretty your face is, what a terrific personality you have, what a good person you are—everything in this world, for women, boils down to body size. Maybe you'll be lucky enough to find one guy who'll like you as you are, and most overweight girls who do will fall in love with that one. But on a singles ski weekend in the Catskills, you'll strike out!

8. Affairs and Affairs

A patient told me that a short time after she had lost a lot of weight, she went into a store owned by an acquaintance. The acquaintance took one look at her and rushed out from behind the counter, hugged her, and cried, "I knew you could do it! I knew you could do it!" She then backed away and said, "When's the bar mitzvah?"

"What bar mitzvah?" my patient asked.

"Well, isn't that what you lost weight for?"

She had the right idea. Many women do come to me to lose weight for a bar mitzvah, a wedding, or a trip. There are usually two problems: they allow a third or less of the time they need to achieve their goal—many women come to me in March to lose 50 pounds for a daughter's June wedding—and they make no effort to stay thin. They don't even wait until the affair is over; that same day they are joyfully stuffing themselves, all dieting forgotten, and the new body to be buried forever in folds of flesh.

I wish more overweight women would be motivated to lose weight for the other kind of affair, but very few men want to start one with a fat woman. There will always be the occasional furtive one-night stands, as long as he doesn't have to exhibit her in public. But no flowers, diamonds, mink, and Caribbean vacations—they go to the thin girls. The overweight girls themselves are partly to blame; they do not send out sexual messages, not because they don't want to, but because they are afraid of rejection. The majority of single overweight women, I'm afraid, have settled for oral gratification, and sacrificed their sex lives to the pleasures of eating.

6.
What Makes a Woman Stop Dieting?

We've talked about what impels women to start a diet; now let's discuss why they fail to see it through.

HUNGER

Right from the beginning of a diet, most of my patients complain they are hungry. They seem to fear hunger so intensely you would almost think they had seriously suffered from it at some point in their lives. Hunger in obesity is very complicated for the following reasons:

1. Mood = Food
This probably is the single most difficult habit to break in the overweight female, because it is so pervasive. Happiness, sadness, boredom, guilt, anxiety, frustration, and nervousness all become reasons to eat. It's a kind of conditioned reflex, and like all such, can be reconditioned.

There are two ways to do this: mood = low calorie food; or mood = no food. Mood = no food is far better. Clean out a closet or write a letter every time you feel hungry. Volunteer for a time-consuming (rather than calorie-consuming) project. Do anything but eat. Next best is to drink diet soda, which falls somewhere between eating and not eating.

If you choose mood = low-calorie food, there is always the chance that you will run out of low-calorie food. Then, unless you've learned control, you will eat something high-caloried.

2. Full = Empty

Being full has no meaning to the overweight. Signals from the stomach have been ignored for so long that when it is full, the brain fails to signal to the mouth to stop eating. Unfortunately, the things that can make us feel hungry are innumerable. They might be certain levels of glucose (sugar) or of insulin in the blood; they might be the sight and smell of food; they might even be the time of day. The only thing we do know for sure is that if you are overweight, you respond inappropriately to food cues. When you are in fact full, you still feel hungry.

This means you have to learn again how to recognize when you are full. You knew it once, but you have forgotten it. Use volume in the stomach as an indicator. Start high; drink eight ounces of fluid, and think yourself full; then reduce it to seven ounces; then to six. Don't be afraid to drink with meals; that helps increase volume. Stop after a certain volume and tell yourself you are full. Repeat this to yourself enough times, and you will be full. The same holds for hunger. Re-learn what real hunger is. Learn it by the clock: 8 o'clock, breakfast; 12 to 1 o'clock, lunch; 6 to 7 o'clock, supper. Anything that you feel between those times is not HUNGER.

Another reason overweight people are always hungry is their high level of carbohydrate intake. Remember the experiment I told you about where people were fed 8,000 calories to try to make them fat? An interesting sidelight of that experiment was the fact that even after consuming 8,000 calories (of which a large proportion was carbohydrates), the volunteers experienced intense hunger. The reasons for this are not completely understood—some say it might be due to low glucose and high insulin levels—but it does show that when the problem is one of distorted food cues, the solution must include a diet low in carbohydrates and fixed-volume feedings eaten at regular hours in order to reestablish accurate hunger signals.

Lately I have been telling my patients with healthy hearts to climb stairs a few times if they are hungry between meals or after supper. Four or five trips up and down an ordinary stairway in a two-story house effectively depress the appetite (as exercise should), increase the pulse rate, and allow the patients to sleep better at night (at least an hour must elapse before they go to sleep, otherwise they are too stimulated).

Finally, may I ask what is so bad about feeling hungry? Pain and itching are far more unpleasant sensations. Hunger *can* be handled, but it must be met in an appropriate way. If you are overweight, the appropriate way is to eat low-calorie foods.

DIETS ARE BORING

The second most common complaint I hear from my patients is that diets are boring—which, correctly interpreted, simply means that a regime of no sweets and no starches is boring. I usually tell my patients that there are three ways of coping with this: (1) The "hard nose" way, (2) the diet club way, and (3) my way.

The "hard nose" philosophy says that diets are meant to be boring! They are not meant to be exciting and fun, they are simply intended to make you lose weight. The more bored you are, the more effective the diet. Get your excitement at mealtime not from the food but from the dinner conversation. Use mealtime for relaxing instead of overeating.

The diet club way of coping with boredom tries to turn dieting into a delicious, delectable experience. Whole recipe books are devoted to low-calorie, wondrous treats, where poached chicken in aspic floats over asparagus with squash soufflé on the side. I feel that the time spent in the kitchen preparing these gourmet meals is too much temptation for overweights. The very act of handling, smelling, and tasting food is probably a major trigger mechanism in overweight eating. Low-calorie *haute cuisine* will only lead to bigger and better things—that is, to more carbohydrates.

I prefer to keep a dieter's food simple. If you want chicken, broil it with salt and paprika, boil it in broth (and give your family the soup), or fix it the way my grandmother did on Friday nights—roast it. If you are one of those women who love to experiment with new recipes (many of us do), then cook for other people, but get someone else in the family to act as the taster—and be aware that you're taking a risk.

My way involves making food as interesting as possible without a lot of fuss. By using a few *simple* food tricks you can keep yourself amused, but it's a delicate balance; once you get too fancy, you could easily revert to the diet club fallacy and become overinvolved with food. Use your imagination. Want something sweet? Do any of the following:

> Eat a sour pickle.
> Eat sauerkraut.
> Eat sweet 'n' sour cabbage.
> Drink diet chocolate milk.
> Drink diet soda.
> Eat stewed tomatoes.

Want Italian Food?

1. American cheese topped with tomato sauce and oregano; fry rapidly = breadless pizza.
2. Lasagna made with ricotta cheese, eggs, and Parmesan cheese, layered with drained hamburger, tomato sauce, and mozzarella cheese = noodleless lasagna.
3. Veal cooked with mushrooms and peppers, and 1 teaspoon oil.
4. Spaghetti sauce—meatless on anything nonfattening (shredded lettuce; broiled eggplant).

Want Chinese food?
It lends itself beautifully to dieting because it uses small amounts of meat and large quantities of low-calorie vegetables. Stir-fry frozen Chinese vegetables using 1 teaspoon of oil; add white meat of chicken, slivered, or a can of tuna. If you prepare it from scratch, use ¼ the amount of oil called for and ½ the cornstarch.
Chocolate?
This is hard to get without calories. One ingenious substitute is chocolate diet soda and powdered skim milk—it tastes just like a chocolate milkshake.
Go for flavor instead of carbohydrates; you can have your tuna fish and grilled cheese sandwich—just leave out the bread.
Diet mayonnaise, mustard, and tarragon make a delicious fat man's Béarnaise sauce.
Use condensed skim milk for crustless quiches and any custard bases. Another delicious crust substitute is broiled eggplant.

Apples and raisins mixed with plain yogurt and refrigerated for several hours makes a dynamite diet dessert.

My way, as you can see, differs from the diet club way in being much quicker and easier. It just uses common sense and clever substitution instead of elaborate fussing in the kitchen.

THE SLIDING SYNDROME

A diet is in its worst trouble when the sliding syndrome appears. Remember when I said that it's the pain-pleasure principle that forces women to diet—when the pain of being fat overcomes the pleasure of eating? But what happens when you have lost some weight, your clothes begin to fit again, your husband starts paying compliments? Then the memory of the psychic pain begins to recede and you lose some of your incentive. You begin to play around. You eat bigger portions. Then you start nibbling on nuts because, after all, nuts are protein. When you start to slide, you never eat anything "bad" at first. Usually you just eat a little more. Then it's the first carbohydrate, and the next day you don't get on the scale. After that you begin breaking your doctor's appointments, and it could be all over.

There's always a month or so after the first appearance of the sliding syndrome before you begin seriously to gain weight. This month is crucial. All your energy must be directed toward restoring the pain you had when you started dieting, or else you must decrease the pleasure of eating.

I would put you on a two-day fast just to break the carbohydrate cycle. Then, to increase pain, I would ask you to do the following;

1. Before you eat something you shouldn't, look at yourself nude in a full-length mirror in good light. It usually is a ghastly sight, and should stop you cold.

2. Buy clothes two sizes too small, and pay a lot for them.

3. Go shopping for clothes just to see yourself in those cruel fitting-room mirrors.

4. Listen to the recording you made the first week of your diet.

Usually this combination will do the trick. Unfortunately, it was during the sliding syndrome that many overweight women developed the dependence on diet pills that led the FDA to restrict them, as I will discuss later. When amphetamines were used in weight control, the dosage had to be increased to get the same appetite-controlling effect they had at the beginning of the diet, but they usually succeeded only in making the patient more edgy, irritable and nervous.

A tremendous amount of psychic energy is required to diet successfully. During the sliding syndrome, that energy must be redoubled.

UNREALISTIC EXPECTATIONS

Unrealistic expectations of weight loss are another frequent cause of failure. I once told a newspaper reporter, "I like my patients to lose weight fast!" The day after the story appeared, an irate patient attacked me: "You lied to me! The newspaper article said that you wanted your patients to lose fast, and I am losing only 2 pounds a week and you told me that was fine." What my patient did not understand was that 2 pounds a week for a female *is* a fast weight loss. The first 2 weeks a dieting female can lose 6 to 12 pounds, depending on the amount of fluid retention she had before starting the diet; after that it's between 1½ and 2½ pounds a week no matter what she does. In any given week she may lose more, but she will drop less the following week to equalize it.

I do feel quite strongly that my patients should be losing 2 pounds a week if it's at all possible. At the rate of a pound a week it will take 30 weeks to lose 30 pounds. Few people have the determination to stay on a weight-loss diet for 30 weeks, but many can easily maintain one for 15 weeks. As for the critics who tell you that the slower you lose it the longer you keep it off, my answer is: Nonsense! Slow weight loss does nothing except keep you dieting longer. You will regain weight rapidly if you go back to your bad eating habits, no matter how slowly you've lost it. That's why I tell my patients, "Diet like hell for as long as you can, as hard as you can." And don't let *any* male, professional or otherwise, hinder your progress by making it harder for you than it has to be. I also tell my patients NEVER, NEVER to compare their rate of weight loss to a man's. Nothing is more discouraging to a female than to see a man losing weight twice as fast by eating more food. As I explained before, that's natural, so don't fret about it. There will also be weeks when you're sure you've dieted perfectly and you don't lose a pound. This is probably attributable to the frequent lag in female weight loss caused by fluid retention. Don't worry; you'll lose the weight the following week. If you don't, then you are simply eating too much.

Remember,
> Hunger
> Boredom
> Sliding Syndrome
> Unrealistic expections

all spell danger to your diet. It takes a lot of psychic energy to stick out a diet, and the kind of dedication most women reserve for being a wife and mother. But if you never let your guard down, no matter what is going on around you, you will come through the experience a thinner person.

7.

Dieting and Our Bodies

In this section, I want to discuss some of the physiological problems that are unique to women in dieting, and help to dispel the mysteries of woman's biology in its complex and subtle relation to weight control.

FLUID RETENTION

It's like the weather—everybody talks about it, but no one really knows what to do about it; everyone is an authority, but no one has any real answers. This is particularly true in dieting, although there is no question that in the female body fat and fluid retention are closely related; in fact, they are almost inexorably bound together.

Beginning in childhood, women have a long history of misunderstanding the whole process of water intake and excretion. It isn't always easy for little girls to urinate. While boys can run to the nearest cover, a girl must interrupt her activities, go home, sit down, and then come back. This is time-consuming and also embarrassing, especially when boys tell her, "We know where *you're* going!" Early in life the female develops a sense of shame about her natural functions; she quickly learns to cross her legs and "hold it in." This lesson is repeated later on in the working world, since women are reluctant to travel past miles of desks to a "john" that always seems interminably far away. And fat women have an extra struggle pulling down their panty hose and girdles and then getting them back up.

What does a woman do in this situation? She learns not to urinate. She learns to ignore the subtle signs that indicate she should use the bathroom. She learns to drink less fluid, particularly water, because water will make her "go" more often, thus throwing away nature's best diuretic. How many times have I heard my patients say, "Water bloats me," or even more commonly, "Doctor, I don't go to the bathroom at all." After being instructed to drink a glass of water, many women immediately report that they are bloated, when they are only registering the volume of fluid in their stomachs. But these feelings frighten them and strengthen their avoidance of drinking water.

True water retention is much more complicated. It results not only from the food you eat and the fluids you drink, but also from the hormones you secrete. It usually only amounts to a maximum of four or five pounds at any given time. But it is a crucial four or five pounds to a dieter, because it can discourage her to the point where she will stop dieting.

When should a woman expect fluid retention during a diet? As everyone knows, a woman retains fluid just before the onset of her period. During that week, she can forget about weight loss, which means that a woman has only three good weeks a month in which to lose weight. During the fourth week, she is lucky if she can hold her own. Even if you have no special problem with fluid retention, nature seems reluctant to let go of any fat during that time.

I'm sure you also know that salt increases the retention of water. So do carbohydrates, even lettuce. But were you aware that alcohol, antibiotics, anti-inflammatory drugs, cortisone, and estrogen also make you retain water?

Lucky males lose weight in a straight slant pattern, like this:

Females lose weight step-fashion, like this:

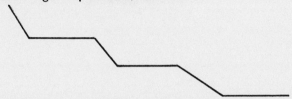

The flat segment represents periods of water retention, or plateaus as they are commonly called, when weight loss stops. Plateaus are dangerous because during them nobody, not even you, believes you are dieting.

If you are weighed during a plateau period, you will become depressed and anxious. For this reason, don't weigh yourself too often: a once-a-week weigh-in is enough. Except perhaps during the first week of dieting, you will not lose enough weight on a daily basis to justify daily weighings. Once a week will tell you if you are headed in the right direction.

And make sure you always use the same scale; they all weigh differently. Most women rely on inaccurate bathroom scales; you should consider investing in a medical (balance) scale if you have any weight problem at all. A good one will last a lifetime, and is a far better investment than exercise equipment, which costs about the same and which you probably won't ever use. However, if you must use a bathroom scale, don't step on it gingerly, jump on it with a vengeance! And try to regulate it with a known weight.

What are some of the solutions to water retention during dieting? First, don't panic. If you have been dieting honestly, you'll ride out the plateau phase. As I've said, many substances can cause a woman to hold water, and if you just wait long enough, the water will eventually be excreted. However, there are some tricks I have found that will keep the water flowing freely and keep dieters happier.

1. *Drink more water.* No, you don't have to drink 8 glasses a day; I find that is a good way to make women stop dieting. Besides making them uncomfortable, a great deal of water on a diet washes out potassium and sodium so that you feel bloated and tired, a terrible combination for dieters. Drink a reasonable amount (4 glasses a day).

2. *Lie down after drinking fluid.* Does that sound strange? Well, women hold more water when in an upright position. Given a test load of 10 glasses of water, a woman will excrete between 5 to 6 glasses after she has been in an upright position, and 8 to 9 glasses if she has been lying down for about 20 minutes. So there is a good reason for a

mid-afternoon or evening rest—with feet up, please. Your husband might enjoy sharing this phase of dieting with you.

3. *Weigh yourself at a fixed time of day*—usually in the morning. Since women have a fluid buildup during the day, even 3 cups of black coffee will register as a weight gain at night. My patients who come into the office in the afternoon invariably say when I put them on the scales, "But I weighed 2 pounds less this morning."

4. *Wear support panty-hose without a girdle.* If you have fat legs, they will trap water like a sponge; so do big, pendulous bellies. A girdle constricts the blood flow from the legs, allowing fluid to form a pool in the lower extremities. (Some women get bladder infections easily, which can be aggravated by wearing nylon, so you may have to wear cotton pants under your panty-hose.)

5. *Go on a no-carbohydrate day.* Eat only meat for two or three meals, and watch that water weight come off. Or go on two no-carbohydrate days. But you must eat only meat or fish protein; whole eggs and cheese won't do.

6. *Have a liquid day.* But be sure you choose the right liquids—no carbohydrates. One of the worst and most common mistakes of dieters is to drink only fruit juices: you might as well eat pure sugar. Women get very hung up with the nutritional value of fruit juices, preferring to forget that they pack a devastating caloric punch. The only fruit juice consumed on a diet should be the juice attached to the fruit that's allowed. (I sometimes make an exception for orange juice on a potassium-low diet—more about that later.)

7. *Find foods (low-calorie, please) that are natural diuretics.* Caffeine, watercress, asparagus, and camomile tea are a few examples.

8. *Avoid salt.* This is my least favorite solution, but most male doctors love it. That's because they're not the ones who have to do without; men rarely need to curtail their salt in order to lose weight. Why should they care that a dull diet becomes duller? But I feel that if I don't enjoy raw vegetables without salt, why should my patients? This seems to me the harshest way to lose water for dieting purposes; I think that if you decrease your total calories and carbohydrates enough, you shouldn't need to cut out salt—*unless* you don't care about it. In that case, by all means try it first.

Most of my female patients, if they are past the age of 25, have at some time or other been given a diuretic. They are now routinely prescribed for bloating, premenstrual tension, swelling of the legs, or high blood pressure. All of these are legitimate reasons for giving diuretics, according to most male doctors; consequently, their use in these situations is rarely criticized. However, when they are used in dieting to make life a little easier for the overweight female, those same male doctors raise a great hue and cry about abuse.

There is no question that diuretics *are* abused—but not by those of us who use them responsibly. They are abused by the gynecologists and internists who write prescriptions that can be refilled indefinitely, or taken at the patient's own discretion. They are abused by women who insist on being rake-thin at the risk of their health.

"Diuretics for a diet? Never!" The medical purists are the first to tell their patients that diuretics deplete body potassium, which can cause weakness, light-headedness, muscle cramps. The implication is clear: if doctors can hang a medical diagnosis on the use of diuretics, they will prescribe them, even if they sometimes cause trouble. But their use in uncomplicated dieting is cause for raised eyebrows.

Hogwash! I refuse to hide behind the medical diagnosis of "periodic cyclic edema" (the gynecologists's favorite term) or the "hypometabolic state" (see next section) to justify treating the very real problem of fluid retention in the dieting female.

Almost all women could benefit from the use of a diuretic at certain times in their dieting program. This does not mean that they cannot diet without it, but only that their weight-loss program will often be longer, more tedious, and unnecessarily frustrating.

When is a diuretic most helpful in a weight-loss program?

1. In women using the birth-control pill or hormones. I love the pill; I think it has given women incredible freedom, but as I've already explained, it wreaks havoc on weight control (among other things). And there is no question that it increases fluid retention in the already watery female. Dieting while on the pill can often be a seesaw affair—one day up in weight, one day down, with no variation in diet. The use of a diuretic here is immensely valuable both physically and psychologically because it counteracts most of the fluid-retaining qualities of the pill.

2. During the second week of dieting. In the first week, there is usually a large fluid loss. Then there occurs what is known as the "rebound phenomenon," in which a woman will actually gain back some of her initial loss. In the first week of dieting the total weight lost is 60 to 70 percent water; the second week, total weight lost is about 40 percent water; thereafter it will be 10 to 15 percent water for the remainder of your diet. A carefully used diuretic will help sustain the initial loss until the fat loss catches up to the water loss.

Unnecessary, some doctors will say. I say, not when you

consider that more women prematurely discontinue their diet during those first two weeks than at any other time. They stop because they want to be rewarded in lost pounds for the effort they've expended. I don't blame them! The "wait-it-out" attitude of male doctors prevails because they won't bother to understand or sympathize with subleties of female eating patterns.

3. After the big weekend. Many people eat or drink more than usual on weekends—not necessarily fattening food, perhaps just larger amounts of diet foods. This can cause the scale to go up 5 pounds even though you have eaten only 2,000 to 3,000 calories in any one day. Since we assume that it takes 3,500 calories to make a pound of fat, that weight gain can't possibly be all fat. The judicious use of a diuretic can show you how much is fat and how much fluid.

4. Period. Period! Diuretics are useful during the week before a woman's period begins, when the scale zooms upward relentlessly. This is a lost week, as I have said, and a time when a diuretic is not only psychologically but physically useful. It is *easiest* during this period to put back weight that you have already lost. In their waterlogged state, the thirsty, newly depleted fat cells can fill themselves up with fat and fluid with amazing rapidity. But don't get into the habit of blaming your period for every weight gain or standstill; only the last four days before your period and the first day of your period are the culprits. I have many patients who insist that they start gaining two weeks before and six days *after* their period. Things are bad enough as they are without exaggerating them.

5. During the dull middle. A diuretic is useful at the midpoint of a diet, when the initial excitement of dieting is over, and the slow, steady weight loss gets slower and slower for no apparent reason. A diuretic often shows you that very day, and not two weeks later, the weight you have lost up till then.

6. For the chronic dieter. Diuretics can help to get the chronic dieter started again when it seems as if her fat cells have become more resistant than ever to breaking down (see my Brand New Re-Start Diet, Chapter 13). There are new findings which show that fat-burning enzymes in the body act more effectively when a dieter is in a dehydrated or "dry" state, as after taking diuretics.

7. In the hypertensive overweight. I find that about 50 percent of my overweight patients have hypertension, or high blood pressure, which will subside when they lose 20 or 30 pounds. Diuretics are useful in keeping their blood pressure down to normal levels while they are

achieving their desired weight; afterwards, the diuretic can often be discontinued. I am amazed at the number of overweight patients who come to me on diuretic medication which has been prescribed for high blood pressure, while no attempt has been made at weight reduction as an adjunct therapy. All of these patients, needless to say, are female.

There is one very important issue that arises with the use of diuretics in dieting: women often have an exaggerated expectation of what water pills can do for them. A woman should understand exactly why she is receiving a diuretic, and what it can and cannot do for her.

A diuretic will not make you lose weight if you are not following a diet!

Don't join the ring-pullers club. A ring-puller is a woman who comes into the office having gained weight from overeating, and immediately starts tugging at her rings: "Look, Doctor, how swollen I am." And it's true, her ring is tight. She tries to blame everything on fluid retention, but let's face it—fingers can get swollen, but they also can get fat. Be honest with yourself; a strong diuretic will not make you lose fat.

Nor will a diuretic make you urinate a flood every day; we are not dealing here with enormous quantities of fluid due to some disease process. A diuretic will help to eliminate most available water immediately, but after the first day you may not even notice the difference. The longer you take a diuretic, the less you will notice the increase in your urinary output.

A strong diuretic will not make you lose weight any faster or better. How many times have I heard women plead, "Doctor, give me a strong water pill." Water pills don't cause weight loss; water pills only help to make weight loss smoother.

A certain amount of water will come back immediately after dieting. The reason for this is that the water you lose initially comes from the glycogen stores in your liver, which hold four times their weight in water; then from your interstitial spaces (spaces between the cells); and last, from the actual burning of fat. When you go off any diet, your liver builds up its glycogen stores again, and the glycogen picks up water. Glycogen is the first line of defense when the body needs energy; fat is the second. The implications of this are that you must overshoot your goal by at least 3 pounds to arrive at your correct weight, because you will pick up again at least 3 pounds of the water you lost at the beginning of your diet.

Increased carbohydrate intake will cancel out any fluid loss, even with a diuretic. I learned this early in my practice, when a woman in her mid-fifties came to me for weight loss. She had a history of heart palpitations, although her heart was not diseased and she was basically healthy, except for her weight. The history of palpitations, however, led me to be quite conservative in treating her. For years her gynecologist had been giving her a diuretic because she complained of feeling bloated. I decided to give her a very ordinary diet and keep her on the same diuretic she had been taking. The following morning I got a frantic call from her husband. "The pills you gave my wife are killing her," he shouted. I rushed over to the house, thinking, "But I didn't give her any pills." When I got there, I realized what had happened. When I decreased her carbohydrate intake by reducing her calories, the diuretic she had been on for years suddenly became very effective and literally washed out all of the potassium in her body, making her feel very faint. After a glass of orange juice she was fine—but off diuretics from then on.

Which brings me to my last point. When any doctor gives any patient a pill, there is a risk involved; what must be weighed are the advantages of the medication against the possible dangers. Diuretics should be used cautiously, but responsible use of them in dieting can mean the difference between success and failure. I make this statement carefully, but I do believe there is a very real connection between fat loss and fluid retention in the female: it's an observable fact that the more waterlogged a woman gets, the less fat she seems to lose. During her periods of greatest water retention, her whole fat-burning system becomes slowed down, and may virtually stop.

THYROID

Every overweight woman wants to believe that she has a thyroid problem, and most secretly believe that's the real reason behind their weight problem, no matter how many doctors have told them otherwise. But what does a "thyroid problem" really mean, and what makes the thyroid so important?

The thyroid gland sits at the base of your neck. You usually can't feel it unless it's slightly enlarged; only a very enlarged thyroid gland (called a goiter) is visible. It's not the gland itself that women are concerned about, however, but the thyroid hormone which it secretes. The "problem" overweight women talk about is an *under*active thyroid gland that is producing too little thyroid hormone.

My patients all seem to understand that thyroid is a very important factor in the breakdown of fat. Since they also realize that they themselves don't break down fat very efficiently, they assume that they must have a thyroid deficiency, or a hypoactive thyroid gland. The thyroid gland can also secrete too much thyroid and be hyperactive. Both hypo- and hyper-active thyroids cause diseases with definite sets of symptoms, in which changes in body weight play only a small part. Most of the time we don't know what makes the thyroid gland get out of kilter and secrete too much or too little hormone. Sometimes it is an infection, or the cause could be stress, while at other times there is no discernible reason at all.

It's easy to see why my patients think they are "glandular" or have a thyroid condition, which to them always means too little thyroid. Look at some of the symptoms of hypothyroidism:

1. fatigue
2. coldness
3. puffiness
4. easy weight gain
5. dry skin
6. brittle hair and nails

Is there an overweight female alive who does not have some if not all of these symptoms? And is it any wonder they are so hopeful that if their thyroid gland is underactive, they can take a magic pill which will make everything all right again, and melt the weight off? All of which is perfectly logical—there's only one catch: most overweight women, probably 95 percent of them, have absolutely nothing wrong with their thyroid function as measured by standard laboratory blood tests. In spite of this, however, all the symptoms listed above do in fact improve with the addition of thyroid. This was first observed in the 1950s, when a condition called the "hypometabolic state" was first defined. The hypometabolic state is a diagnosis made when all the symptoms listed above are present, but when the thyroid blood test is normal. When a certain type of thyroid (Cytomel) is given, all the symptoms improve and weight loss is facilitated. Most doctors do not believe that there is such a disease entity as the hypometabolic state. But I do!

What are we going to do with these poor women who eat so little, gain so much, feel lousy, and are convinced that thyroid will help them feel and look better? The purists tell us, "You never should give thyroid unless there's a thyroid deficiency." Or, "There are inherent risks in giving thyroid medication." Or, "Giving thyroid doesn't even increase fat loss; when you give thyroid, you increase protein loss."

This is the dilemma in which I find myself: I want to help overweight women lose weight more easily. Thyroid helps them to lose weight because thyroid helps burn fat. But the majority of overweight women do not actually *need* thyroid—meaning that the standard tests reveal no deficiency.

I have watched the thyroid picture evolve from the old basal metabolism test, which resulted in almost everybody being put on replacement thyroid, to the thyroid blood tests, which have almost nobody on thyroid, to the newer photomotogram (ankle reflex test), which is starting to put women back on thyroid again. Taking all this into account, my own attitude is that obesity is unquestionably in part a metabolic problem. Something in the body of the overweight female isn't burning fat the way it should and that something could be thyroid. Furthermore, I fail to find thyroid to be as dangerous a drug as the purists would have us believe. Of the hundreds of women who come into my office already taking thyroid (for various reasons), I have yet to see a serious complication resulting from the drug when it is used in the proper dosage by a capable physician.

I find thyroid very helpful in weight loss, but I use it only (combined, of course, with diet) in cases of:

1. *true thyroid deficiency*
2. *resistant obesity—women whose calorie counts have gone down to below 600, and who are still losing only ¼ to ½ pound per week*
3. *postsurgery (for one year after surgery weight loss in obese females is very slow)*
4. *postpartum (women don't lose weight as efficiently up to six months after the birth of a baby—unless they are nursing)*

I have found that the addition of as little as 25 micrograms of Cytomel (equivalent to 1 grain of thyroid) will allow previously resistant patients to lose 2 pounds a week on my basic Core Diet, and I see no reason not to prescribe it in most cases. But why is the hypometabolic state not taken into account by more doctors? I think we are misled by the techniques we have to measure thyroid, which are inadequate for clinical use. They measure only the quantity of thyroid hormone circulating in the blood and not how it *interacts* with the cell to break down fat. I am sure the defect or deficiency is at this level of cell interaction.

The thyroid story is still changing, but it is beginning to look as if the old BMR (Basal Metabolism Rate) test, which showed that most overweight women were *hypo*thyroid, might have been right after all!

AMPHETAMINES (DIET PILLS)

The controversy about diet pills continues to rage. A few weeks ago I picked up the paper and read a report of a panel of five male physicians—four pediatricians and a psychiatrist—who are fighting to have diet pills taken off the market on the grounds that they are dangerous and ineffective.

"The world needs the birth-control pill," said one of them, "but I am unable to identify a similar need for amphetamines and related drugs."

You are right, Doctor; and by your line of reasoning, the world does not need sleeping pills either. People overdose on them, kids abuse them, and what difference does it really make if you get a few hours less sleep at night? Nor does the world need the minor tranquilizers, such as Valium and Librium—even though I'll bet more prescriptions are written for them than for any other drug in the world. We can certainly all tolerate a little anxiety, particularly when it's women who are doing the tolerating. And of course, Mr. *Male* Doctor, we'd expect you to tell us the world needs a birth-control pill when it's not you suffering from the blood clots, the falling hair, the depression. All of these drugs are imperfect ways of reducing human misery, but they do offer some help. Birth control *can* be accomplished without a pill, just as dieting can be accomplished without a pill, but it's a lot more difficult.

Amphetamines were devised originally to cure narcolepsy, a condition in which people would suddenly fall asleep in the midst of normal activities, such as talking or driving; later they became popular as antidepressants, to pep people up and get them moving. After specific antidepressants were developed, it was no longer necessary to prescribe amphetamines for this purpose. But since it had been noticed that they also curb appetite, doctors began dispensing them to dieters; and they gave them out wildly, writing prescriptions for six months at a time to anyone who asked for them, or ladling them out by the handful from large vats in their offices. There was no discrimination as to who got the pills, and what was more serious, there was no follow-up. Whether or not a patient lost weight didn't even seem to matter because, unfortunately, amphetamines lose their capacity for appetite control long before they lose their ability to give energy. Male doctors kept giving their female patients increasingly large doses of amphetamines to control their appetites. And then, about ten years ago, the worst mistake of all: deadly drugs out of Texas, distributed by irresponsible physicians, combined amphetamines with digitalis, giving them a still worse name.

At last the Food and Drug Administration began to tighten controls on amphetamines, and the drug manufacturers began looking for drugs that would have the same ability to control appetite, without the stimulating properties. They eventually succeeded in producing a group of amphetaminelike drugs that did not send patients climbing the walls; that curb appetite but are not chemically related in any way to amphetamines.

But there are still some questions remaining about the safety of diet pills.

Do amphetamines or related drugs cause high blood pressure? They could aggravate *existing* high blood pressure. Obesity can cause it, and many obese people have taken diet pills. Which came first? Probably the obesity, then the high blood pressure, and, only last, the occasional diet pill.

Do amphetamines or related drugs aggravate heart trouble? Probably. But who would prescribe amphetamines (or most other drugs, for that matter) to people with heart trouble? Obesity itself aggravates heart trouble. I feel that it is safer for an overweight person with heart trouble to use a mild diet pill than it is for him to carry around 50 to 100 pounds of excess weight. And many very obese women feel that they psychologically need a diet pill to get started on a diet.

Are amphetamines or related drugs potentially addicting? Any drug that affects mood and state of consciousness except, perhaps, for the major tranquilizers (like Thorazine) appears to have the potential to become addicting. But are heavy ladies who take these drugs in order to diet likely to become addicted? Practically never—they're much too fond of food.

Amphetamine addicts are usually teenagers who want to get a thrill by taking six or seven times the normal dose of the pill; people working in jobs that demand so much from them physically and emotionally that they need the pill for pep; or people who would get addicted to anything. Are these the kind of people you see in the office of a doctor who deals in obesity? Rarely! You see mostly people who would rather eat than take pills, people who are sensitive to the slightest stimulation the pill has to offer. Obese women are naturally phlegmatic. That which gives them a burst of pep or stimulates them, they do not find desirable. In fact, even small amounts of excitation frighten them. Male physicians in their greediness have created a problem with diet pills—they've given them out too freely in the past; now guilt at their former excesses has caused them to over-react. They should campaign instead against cigarettes and hand guns—they're far more dangerous.

Diet pills are only a small slice of the dieting pie, but they can be a useful part. They work directly, in a way that is not yet completely understood, upon the appetite centers in the hypothalamus, and also probably stimulate the release of norepinephrine (which gives an effect of satiety). The new diet pills have virtually no stimulating effect and thus do not lend themselves to abuse. But in any case, it's the responsibility of the physician to monitor any medication he prescribes; he must know his patient and know whether the pill, if he decides to use it, is being used to good advantage. *There is no sense taking a diet pill if you are not on a diet program,* or if you do not intend to cut down drastically on your food intake.

I can practice medicine without recourse to diet pills, but I hope I don't have to. They are extremely useful for dieting women who have to cut their calories down to such a low level that it becomes a real hardship. Diet pills are especially useful the first eight to ten weeks of a diet. (After that, their ability to control appetite is considerably reduced. They should then be discontinued for several weeks and resumed when the body has once again lost its tolerance for them.) They are useful also at those unavoidable times when temptations are great, and at the end of a diet, to help get rid of those last, resistant ten pounds.

At the moment, diet pills are out of fashion. Women are afraid of them, afraid of becoming addicted. And of course it's the male physician who lays on the fear and the guilt, accusing those who do take them of being somehow weak and inferior (ten years ago, when the winds of fashion blew the other way, we never heard this line). Of course diet pills aren't the whole answer, but neither is any other program I have yet seen. There's no one simple answer.

However, a judiciously used diet pill *can* curb your appetite and focus your attention on your diet. Any pills can produce side effects such as dry mouth, rapid pulse, and irritability; these disappear as soon as the drug is discontinued. Even the mildest pills can make *some* women nervous; the more neurotic the female, the more sensitive she will be to the effects of the diet pill, because she is so tuned in to her inner feelings.

Dieting is terribly difficult for many women, and for some it's almost impossible. Diet pills can make it easier for them to reinforce their desire to curtail their food intake in the face of overwhelming temptations, and I can see no reason to deny them this help.

MOOD

I mentioned the effect of diet pills on the mood of neurotic, and therefore more sensitive, women; but definite mood changes occur in dieting whether or not the dieter is using pills. I admit that if she is on medication, she may become irritable, but there are other reasons for irritability as well:

1. The dieter invariably feels sorry for herself and expects everyone else to pity her, too. If her family and friends persist in treating her as they always have, this is annoying to her.

2. As I said before, I'm sure that in many cases food acts as a tranquilizer; when you deprive a woman of her tranquilizer by putting her on a diet, you have problems.

I've had husbands call me to change their wives' diet pill because it made them so cranky, when they weren't taking diet pills at all. I advise my overweight patients that they have two choices: since food is a balm for irritability and dieting is depriving them of this balm, they can choose to be either elephantine or edgy, at least until the end of the diet. For this nonspecific kind of irritability, I find that a few Tylenol or aspirin help.

FATIGUE

"I feel so tired," "I have no energy," "I'm all washed out": these are among the complaints most commonly heard from women who visit a doctor's office, and my practice is no exception. Almost all my patients seem to suffer from chronic fatigue, whether before, during, or after a diet. I used to think that obese women were tired because they had so much extra weight to carry around. But if that were true, the heavier the woman, the more tired she should feel; it would then follow that the more weight she lost, the more energetic she should become. Neither of these assumptions is borne out by the facts. The amount of fat seems to bear no relation to the amount of fatigue, and many women still lack energy after a diet. Furthermore, although overweight women carry twice as much weight, they often do half as much work, so that the energy they expend amounts only to that expended by women of lesser, normal weight. Fatigue is therefore not necessarily a direct consequence of overweight alone. By all measurable indices, the body of an overweight female is not working harder.

I feel that a low level of energy, or what is interpreted as fatigue, is characteristic of the hyperplastic (too many fat cells) overweight's body type, as well as her body build. Most chronically overweight women are endomorphs, meaning that they are inclined to be soft-fleshed, sensuous, and rather indolent. Fat or thin, they tend to be a sluggish lot. Some of the low energy levels I see can also be attributed to a high carbohydrate intake, at least in the fat, nondieting state. Therefore, I believe that low energy levels are built into the systems of most overweight females. However, when 90 percent of my patients complain of feeling tired or having little energy during the course of a diet, I am compelled to look for other phsyical and/or emotional reasons as well. If I fail to find either of these, I conclude that they must simply be considered genetically tired people.

What are some of the real or imagined reasons for being tired during the course of a diet?

One possibility is the hypometabolic state I described in the section on thyroid. Another is:

Low Blood Presure
If you get refused by the bloodmobile when you are dieting, more often than not the reason is low blood pressure. In order to be defined as low, the systolic pressure, or top number, should be 100mg or lower, and the diastolic pressure, or bottom number, should be 60mg or lower. There is absolutely no reason why low blood pressure should make you feel tired—unless you are in shock, and shock is not a chronic state! Blood pressure measures the force required for your heart to pump your blood through your arteries, and also the elasticity of your arteries. If your heart doesn't have to work as hard to get the blood around, that's healthy. If your arteries have a lot of elasticity, that's healthy, too. Low blood pressure, then, is a good thing.

So why, you ask, won't they take blood from you at the bloodmobile? Because low blood pressure does present one problem: if it falls below a certain level, either by reducing blood volume (as in taking out blood) or failure to adapt to a sudden change in position (as in standing up from a sitting position), you may experience dizziness or light-headedness, and you might feel faint. However, the minute you put your head down the brain gets its blood supply back again and you usually do *not* lose consciousness, although until then you may think you are going to. Obviously, the bloodmobile personnel do not want to have to pick you up off the floor if this happens. It doesn't look good to the other blood donors, and besides, you could injure yourself in falling.

Females are very likely to experience light-headedness while they are on a diet, but they should not confuse this with real dizziness. This light-headedness is sometimes described as a spaced-out feeling or a feeling of faintness, and is usually due to the lack of rapid adjustment of the blood vessels when you change position (for instance, from sitting to standing), which allows blood to pool in the lower extremities and momentarily to leave the brain. This *rarely* causes fainting. Many of my patients say to me, "Boy, I almost fainted, but I stopped myself." If you were really going to faint, you could not consciously stop the process—you would have no control over it.

You *can* control sudden hypotensive (low blood pressure) light-headedness by changing positions slowly (from sitting to standing). If you get a hypotensive episode, you can stop it by lowering your head between your knees or lying down for a few minutes.

Anemia
Anemia, also called "low blood," is, rather than thyroid, actually the deficiency I most commonly see in overweight women. Anemia is a term used to designate the failure of the body to produce enough red cells, or a failure to incorporate enough iron into the red blood cells being produced. Occasionally it is due to an obvious cause such as heavy menstrual flow, bleeding hemorrhoids, or the recent delivery of a baby; or it may result from a natural preference for high-carbohydrate foods and a distaste for food rich in iron, such as lean meats, dark leafy vegetables, and organ meats. Sometimes the anemia originates in infancy, when the well-meaning mother feeds her child an excess of milk over iron-rich food in order to ensure adequate calories. Anemia in the obese may also be caused by an excess amount of cellular and water mass relative to the existing volume of blood cells, which dilutes the effect of a normally adequate number of red blood cells.

Cases of iron deficiency have become less common now that so many high-carbohydrate foods like cereals and bread are fortified with iron, and also because so many of the multivitamin supplements have iron in them. But while it is true that we can walk around with an extremely low red blood count and still function, even mild anemias can produce profound fatigue in overweight females. The fatigue is partly a manifestation of decreased oxygen supply to the cells, and is reversible (in the absence of underlying disease) by taking iron tablets. But remember, the diagnosis of anemia can be made only by a blood count, and should not be made on the basis of color, or creases in your hands, or the paleness of the conjunctivas of your eyes. Iron supplements do have a tendency to change bowel habits; most people find them constipating, but in 10 percent of patients they will cause diarrhea.

Serious cases of anemia (below 10 Hgb. and 30 Hct.) are rarely nutritional in origin (unless you have been starving yourself for a long period of time), and should be evaluated by a complete physical examination.

Medical Hypoglycemia (Low Blood Sugar)

This is the single cause of fatigue that doctors most frequently look for today and rarely find. The press has done an incredible job of promoting hypoglycemia as the all-purpose cause of every human ill, so much so that women don't even mind taking the long, costly, five- or six-hour glucose tolerance test required to prove it; a great many doctors now even incorporate it into their routine workup. Hypoglycemia has become to the seventies what hypothryroidism was to the fifties and anemia to the sixties.

Hypoglycemia is really a symptom, not a disease in itself. It can be severe or mild, depending on the disease that is causing it; it may also have no cause, in which case it is called functional hypoglycemia. In essence, when the blood sugar dips below 50 mg percent certain symptoms may occur. These symptoms are probably brought on by a release of adrenaline, which causes anxiety, pallor, shakiness, sweating, palpitation, and a subsequent compensating rise in blood sugar. Needless to say, this condition is acute, uncomfortable, and frightening. But recently, any chronic low-grade, nonspecific, vague case of fatigue and depression is being diagnosed as hypoglycemia.

Since many women have these symptoms, they believe that they are hypoglycemic. So I find myself once again practicing defensive medicine and ordering a glucose tolerance test—not to rule out diabetes, which is what I formerly used the test for, but to rule out hypoglycemia. I find very few true hypoglycemics, but I do make some interesting observations during a glucose tolerance test. When you take a glucose tolerance test you are given a fixed amount of sugar water to drink, after which your blood and urine are collected at hourly intervals. We look for the following:

1. How high the sugar goes and when;
2. Whether or not you spill sugar into your urine;
3. How low the blood sugar drops and when.

When the blood sugar drops to a certain level, but *not* to the critical level required for the diagnosis of hypoglycemia, emotional symptoms *do* occur. They are not psychological, because the patient has no idea that her blood sugar is in fact at its lowest point. Some patients get very anxious, some burst into tears, some feel very shaky, some become very confused. All these things occur while their blood sugar is at its lowest, but not necessarily hypoglycemic, levels. Therefore, I think we should recognize that a woman can have a *relative* degree of hypoglycemia—relative to her own rate or degree of decline in blood sugar. I have come to the conclusion that blood sugar levels probably do have a major effect on mood in females, perhaps even as much as the mass media would have us believe.

The treatment for both true and relative hypoglycemia is a high-protein diet with frequent small feedings. Since this is also an excellent way to reduce your weight, it can easily be worked into a diet program. Remember, sugar is *not* the answer for hypoglycemia. It will send your blood sugar up high, but afterwards it will plunge again to new lows.

Pregnancy
Early pregnancy is another possible cause of fatigue in a woman on a diet. This is often an elusive symptom, especially when many overweight women have irregular periods; but if pregnancy is a possibility, it should always be looked for first.

The Birth-Control Pill
The birth-control pill can cause fatigue in many females, but usually only in those who have been taking these pills for a long period of time. The reason for this is unknown.

Low-Potassium State
Potassium is a chemical element that is vital to the strength and well-being of all the muscles, including the heart. Together with

sodium, it also helps regulate internal body fluid balance and, along with oxygen and glucose, it plays a part in other important metabolic activities. At any time during a diet, water may be lost in excess, and whenever water is excreted, so is potassium. If you take a diuretic, more potassium may be lost. Women vary in their sensitivities to the loss of potassium; some don't feel it at all, while others become so weak they cannot get out of bed.

When you are dieting and in a low-potassium state, you may experience fatigue, weakness and/or muscle cramps. Hypokalemia (low blood potassium) can be diagnosed only in a laboratory. However, if you think your fatigue is due to potassium depletion, an eight-ounce glass of orange juice (almost 150 calories, unfortunately) will help you feel better in about twenty minutes. Remember, it is the *potassium* in the orange juice and not the sugar that makes you feel better, in cases of potassium depletion. Since orange juice is also a quick source of sugar, however, it is taken by diabetics who have had too much insulin in order to replenish their sugar rapidly.

Many women believe that when they feel weak during dieting, it is due to a low-sugar problem, but most of the time it is due to low potassium. Orange juice, bananas, tomato juice (not as good a source), white meat of chicken, and spinach all replace potassium. Dried fruits are rich in potassium, too, but they are also rich in calories.

Layman's Low Blood Sugar
The conviction held by so many dieting women that any fatigue they may feel is the result of low blood sugar is, I'm sorry to say, only wishful thinking. The cure for fatigue is *not* to consume sugar; minor fluctuations of normal blood sugar should not cause tiredness or fatigue. It is a pity so many eating binges are triggered off by the belief that a momentary weakness is actually due to a need for sugar. I always tell my patients, "Look, if you were 90 pounds dripping wet, then you might need sugar for energy, but with your fat reserves, sugar is the last thing you need!"

One research team postulated that the patient's desire for something sweet could be satisfied by a certain hard candy they manufactured which contained only 10 calories and was made from sugar, caffeine, and a local anesthetic. They felt that the sugar would satisfy psychological needs, while the caffeine would give the patient a little boost, and the local anesthetic would dull the taste buds. The patients were instructed to eat these candies whenever they felt hungry. The results of the study showed that groups using the

10-calorie lozenges dieted as successfully as groups who used appetite-suppressant drugs.

I was very excited at the possibility that here at last was a solution to the dieter's perennial problem of craving for sweets. I sent my patients out to buy cough drops made with the same ingredients, and instructed them to take one cough drop with a cup of coffee (for the caffeine). I was disappointed to find that this did not in any way help the dieters control their cravings; on the contrary, it stimulated them to eat still more sweet foods.

Vitamin Deficiency

It is not my purpose here to discuss vitamins in detail, except to say they are organic substances, occurring naturally in nature, which are essential to the normal growth, development, and functioning of the human body. I used to assume that anyone who ate a so-called balanced diet got enough vitamins. Recently, however, since many of my patients have been asking whether their lack of energy is caused by a vitamin deficiency, and wondering whether they should take supplements in addition to their regular diet, I have examined the subject of vitamins more carefully.

The FDA has decreed what quantities of which vitamins we need on a daily basis. The fat-soluble vitamins—A, D, E, K—can be stored in body fat and therefore can be toxic in large doses, but vitamin C and water-soluble vitamin B and all its complexes are not stored and need daily replacement. How much replacement depends on whether or not you listen to the FDA. I tend to go along with the more radical fringe in this area who feel that the larger the amount, the better the effects.

Let's talk first about vitamin C. In my experience, high doses (1,000 mg. daily) of vitamin C seem to decrease the number, duration, and severity of colds in my female patients. I have also seen mild depressions helped by large doses of vitamin C, although I have not seen vitamin C give more energy; the feeling of well-being is much more emotional than physical.

The B vitamins are probably the most interesting vitamins of all; I suspect that we have not even scratched the surface of their value. If I give a basically well patient a shot of B_{12} when she feels fatigued, and twenty-four hours later she feels a great deal better for about a week —and this happens to 90 percent of my patients who receive B_{12} —I refuse to believe that this is all "in her head," or the result of "the placebo effect."

For years family doctors have been taking it on the chin for giving their patients B_{12} shots, which the purists say do nothing. It's not true. Far too many male internists pay no attention to how people *feel*. They are very suspicious of anything that makes people feel better, if they can't explain it with test-tube logic. Empirical evidence is accumulating that C and B vitamins make people feel better, although we are not sure how or why. They probably act through various enzyme systems in ways we have yet to discover. I for one am delighted that these vitamins give my patients a sense of well-being, whatever the reason; I don't get uptight simply because I don't know why. The vitamin story has yet to be written, and I think when it finally is, it will be a very exciting one!

Incidentally, it is not true that vitamins increase your appetite! That is a myth. They make you feel better, though, and some overweight women think feeling better is the same as being hungry.

Depression
Depression and/or boredom are usually the last reasons, I consider as causes of fatigue in the dieting female. This is not because they are not very prevalent, but because all psychiatric diagnoses have to follow a process of *exclusion*. In other words, only after you have excluded all the physical reasons for a symptom such as fatigue may you then consider the possible psychological reasons. Often depression and boredom decrease as the patient loses weight. The depression that I see in the overweight female is *not* the classical depression which causes loss of appetite, but rather a mild, chronic self-hatred or self-pity. Usually no medication is needed to help the patient through this state; her diet and weight loss will work wonders as a cure.

If you are dieting you may experience some fatigue. I'm virtually certain that the reason will be found to be among those listed here. Now that you understand them, don't allow these minor problems to stop you. Use your new knowledge to help you stay on the right track!

EXERCISE

One of the problems of overweight females most resistant to change is their reluctance to move their bodies. If you have a tendency toward overweight, exercise alone is not the answer, since it has been shown that in the obese, exercise without restricting food intake does not result in a decrease of body fat. But it does play an important part in a successful diet program.

Exercise is probably most important for heavy teenagers and for women over 50; these two groups move a great deal less than their thin counterparts. I used to despair when older women patients would walk into my office (and I mean "older" in a fat-burning sense; in my practice, I define middle age as beginning at 25). Even on a severely restricted diet they could lose only between ¼ and ½ pound per week; faced with such discouraging progress, they would invariably give up.

A solution to this problem occurred to me when a patient I had been working with for several months decided to go to Duke University to participate in their famous rice diet program. Its Spartan discipline required her to eat a limited amount of rice and fruit and to walk about five miles per day. When she came back from Duke about five months later and visited my office, I could scarcely believe my eyes. She had lost nearly 40 pounds. When I checked her dietary intake, I found that not only was it high in carbohydrate, it was not much lower in calories than she had been eating on my diet. What made the difference was the daily five-mile walk. She told me that when she was instructed to do this, she said to the doctors, "I can't do it, I'll die."

"Just don't drop dead on the street," one of them told her, "it would disturb the traffic."

I can assure you that at home this lady could not walk across the street without puffing and panting. Her weight loss with me had been ½ to 1 pound a week on a diet as strict as she could tolerate. It was an impasse, because at that rate it would have taken her a year and a half to lose the weight she needed to lose, and she could never have stayed on such a strict diet that long.

I was so impressed both with the lady's improved physical condition and her weight loss that I decided to start all my female patients on a walking program and to push it hardest with the over-50 dieter. "If you follow this diet and walk one mile a day, you will lose weight the way you did when you were thirty," I promised my older patients, and it worked every time!

One of the saddest things I see in my practice is the overweight female teenager. She is often alienated from her peer group to begin with, and if she is trying to lose weight, her thin friends will sometimes pressure her into eating fattening food. Often the only way she can "go along with the crowd" is by conforming to their eating habits, since she isn't given the chance to take part in their other social rituals such as dating. The irony of this situation is that the overweight teenager may eat significantly fewer calories than girls her age of normal weight. Though I feel her basic weight problem is her inborn inefficiency in converting food to energy, studies have shown she is usually less physically active than her thin peers. Researchers who took pictures of girls swimming found that overweight girls spent the same amount of time in the water as thin girls, but most of that time they were just standing around. One obvious solution is to get the overweight female teenager moving, in addition to cutting her calories; exercise would seem, therefore, to be the great equalizer in the adolescent no less than in the older female.

What about the exercise pattern of overweight women between 20 and 50? Many of my patients earnestly tell me that they have joined a health spa and are going religiously five days a week, but it always turns out that they neither exercise for a sustained period of time nor increase their pulse rate—sometimes they don't even perspire. Don't get me wrong. Health spas are great for tightening up muscles and building stamina, but *not* for the sustained burning of calories—at least not the way my patients use them.

American women have not been taught to enjoy using their bodies in strenuous exercise. They are brought up to believe that their bodies should be objects for sexual and aesthetic appreciation, but they do not see them as strong, vital organisms. As children we are not taught to hike, or to climb mountains. Walking in the woods is considered too dangerous and unladylike. When we get older, walking in the rain ruins our hair or the sun dries our skin. Creatures who are meant to be ornaments can't move.

When I talk about physical exercise, I'm not talking about fashionable calisthenics, but about getting yourself moving for the sake of breathing hard, feeling your blood pump rapidly, and working up a sweat. European women know how to move, and they appreciate their bodies in action, perhaps because they have a greater appreciation of their own sexuality. American women with good bodies are like mannequins, to be looked at, admired, and placed on a shelf. European women enjoy the fact that their bodies are functional and not just ornamental.

However, the benefits of exercise have more to do with well-being and energy than with weight loss; I've never seen a patient of mine lose weight by exercise alone, although I *have* seen men do it. Evidently women don't exercise long enough, hard enough, or fast enough.

My patients become very concerned about the possibility of creating *more muscle*, which, they solemnly tell me, weighs more than fat. (Women love to believe this fairy tale.) I have never seen this happen. It's hard to build muscle—wrestlers and football players are fed special high-protein diets and exercise constantly to build up muscle mass. No matter how hard they try, women simply cannot build up muscle to that degree. The best an adult dieting female can hope to do is firm up and tone up existing muscles so they will take up the slack left by the departed fat. And if the skin cooperates, it will fit snugly over the deflated area, resulting in a change of body measurements.

"But exercise makes me so hungry," is another favorite excuse I hear from patients. We are all primed to think that exercise makes us hungry. When we look at TV or read a magazine we see sleek, beautiful people eating and drinking and proclaiming, "Wow, nothing tastes better after a hard run on the slopes than hot chocolate," or "Off the surfboard and into a delicious hamburger." Naturally, dieters who exercise invariably feel that they are starving when they return from the slopes, rink, pool, or hike, but this also has a lot to do with their constant tendency to equate feeling good with feeling hungry. So they end up eating all the calories they lost, and then some.

Actually, most exercise programs will tell you not to eat for about an hour after exercising. The reason for this is to give your oxygen supply a chance to stay where it is needed, and not get diverted to your gut. In reality, *exercise decreases appetite;* the waste products of muscular activity after exercise actually inhibit appetite. If it weren't for all the misleading propaganda, exercise would be a perfect method for *controlling* hunger, rather then creating it.

Can you selectively change measurements by exercise? Another pipe dream. You will lose equally all over, no matter how hard you pound that hip against the wall. The pie-in-the-sky idea is that if you rub, stretch, pound, or massage a certain bulging or flabby spot, you will break down the fat in that area and it will somehow miraculously be consumed. This is nonsense. You would be in a sorry pickle if you could go around breaking down fat cells; the little devils could be very dangerous, and end up clogging the blood vessels of your heart and head. What I am saying is that you lose equally all over, not in proportion to what you think is needed.

The Best Exercises for Females

Exercise charts tell us how many calories are burned by various activities, but when I look at these charts, I doubt that my patients will burn one-half that amount doing the activities listed. Except for light housekeeping, these charts seem to have been devised with men in mind. Men (and thin women) are heat-producing animals: they feel warm, they sweat, and they give off heat. Heavy females, on the other hand, except when they are having hot flashes, are cold (even though the only use fat has, besides storing energy, is to act as insulation), and many perspire only lightly, if at all. They do not give off heat when they exercise at the same rate as men and as their thin sisters. I have patients who play tennis several hours a day, which is supposed to burn 420 calories per hour. This, plus what I am feeding them, means they should be taking in zero calories per day, but they certainly don't lose weight at the rate of zero calories per day.

Personally, I think walking is the best exercise for overweight women because:

1. You don't need any special equipment.
2. You don't have to join anything to do it.
3. Walking increases physical fitness as well as promoting weight loss (*with diet!*) Women can become thin and still be very unfit. Although I deal primarily with weight loss, it is much better if my patients become physically fit along the way.
4. Most people can walk, and it can be performed at different levels of energy expenditure. (You can work up later to jogging and running, if you want to.)
5. Walking can be done anywhere, any time.

Walking is the one exercise in which you are solely responsible for carrying your own weight from one point to another, which means that the heavier you are, the more calories you burn moving yourself. For once, you can actually make your fat work for you. In bicycling, the wheels of your bike relieve you of your weight and make it easier to go from one place to another; in swimming, the water buoys you up. In walking, however, the weight is all yours, and the exercise, combined with air currents and the friction of the road, gives the fat person a good workout. You should walk for at least twenty minutes every day at the rate of three miles per hour. *This* exercise regime, combined with a personalized diet, is all I usually ask my patients to follow.

But it's surprising how much trouble some of them have understanding me when I talk about walking as exercise. They begin by asking, "Doctor, do you recommend any exercise?" followed by, "How about bike riding?" or, "I just joined a health spa. We do calisthenics for an hour." When I've finally got my point across, they will answer, "I don't have time to walk," or, "I'm too tired. I work hard all day," or, "It's too dark to walk when I get home," or, "I live in the country. I can't possibly walk on those dark roads." Or else, "I live in the city. It's too dangerous to walk." Some patients are more inventive and tell me: "I walk all day at work. I walk *miles* at work," or, "It's too cold, hot, rainy, windy, polluted, mosquitoey," or even, "There are no sidewalks where I live." Physical reasons such as "I have a bad knee (back, ankle)" are also high on the list of excuses not to walk. After a while, even I begin to think I am suggesting something illegal, immoral, or dangerous!

Why do you suppose women, particularly overweight women, resist the whole concept of walking as a form of exercise when it's so easy, cheap, and readily available? First, walking has a poor press; nobody can make money from it. You never see ads in magazines telling you to "lose inches by walking." Second, walking alone may seem boring, especially when you are going for distance; it's better to have a companion. Someone should start a fashion for teams: "Okay, Joan, Ina, Peggy, you will walk with Harriet today, and tomorrow we will switch." Or round robin walks—the winner will walk with Amy. Third, walking is concentrated physical work and most fat women *avoid* concentrated physical work. What work they do, they like to spread out. Fourth, it's hard to cheat on a long walk; there's usually no place to rest, have a cup of coffee, watch TV. And there's always the knowledge that once you have arrived at your destination, you must return the same way you got there, on your own two feet. Finally, overweight people do not enjoy being out of breath or having their heart pound over 70 beats per minute. But as much as my patients resist me, I still say a good diet, plus at least a twenty-minute walk daily, is the key to the most efficient weight loss.

I used to recommend jumping rope as an exercise for women, because there I had to plead for only *two minutes* of time. Jumping rope is very strenuous exercise which moves every part of your body, but it can also exert a downward pull on the ligaments of your breasts and uterus. Therefore I now only recommend this activity to those who are very young and have very firm, tight muscles.

Many women find the totally imaginary concepts of hard fat and soft fat very appealing. Often they will punch their leg and say, "I'm going to have a hard time losing this, I'm so solid," or they will seriously tell you, "This should be so easy to lose, I'm so flabby." There is no such thing as hard or soft fat. There is firm muscle and firm skin, flabby muscle and flabby skin, and just plain fat. However, I do believe that there is new fat and old fat. They look the same outwardly; the difference is that new fat seems to be less resistant to metabolic breakdown because the cell walls are permeable and more amenable to entrance by the enzymes which help mobilize them. Enzymes have entered old fat too many times; the cell membrane has built up a resistance to being broken down. I mention this because, on the whole, *fat is fat.* If it has become resistant, the reason is its age or cell size, not its position in the body. Fat cannot be broken down in the body, but it *can* be used for energy if energy is not forthcoming from some other source. (This is the basic principle of dieting.) *Combined with reduced calories,* consistent, aerobic exercise *can* produce slightly faster weight loss.

But where exercise really counts is in the *maintenance* of weight loss. A woman is bound to relax when she goes on a maintenance diet, and will tend to eat more of even non-fattening food. Overweight females, as I've mentioned more than once, can never eat as much as they want of anything except vegetables like lettuce and cucumbers. However, with the addition of a realistic exercise program it is possible to eat more food without gaining weight.

MENSTRUATION

From the moment a female starts having periods until the day she dies, cyclical changes are constantly occurring in her body. Because we are a medically oriented society, we have given these cycles scientific names. When we put out an egg for fertilization, we call it ovulation; our uterus bleeds every month when there's no baby, so we menstruate. When our uterus fills up with a fetus, we're gravid; when it empties, we're postpartum. When our periods disappear, we have a change of life, or menopause. If we swell before periods, we are labeled premenstrual.

Sometimes, as women, we seem to be nothing more than a collection of medical conditions that change according to the balance or imbalance of our female hormones, estrogen and progesterone. Now that the male medical world has categorized every one of our bodily states, they can sit back and tell us that each phase we're in at the moment is either biological or psychological. Well, it's neither; it's physiological, especially when it's a question of weight and overweight. The whole natural cycle that we are supposed to "grin and bear" wreaks havoc with weight, diet, and appetite, producing changes which are too regular and predictable to be written off as simply figments "of our imaginations."

For instance, when you are premenstrual you are more irritable, hungrier, 5 to 6 pounds heavier from water retention, unable to lose weight, more susceptible to weight gain, crave sweets (especially chocolate), and depressed. Worse, if you are overweight, along with all your other psychological hangups, the sequence depression = mood = food takes over. When you are overweight and premenstrual, you hold still more water, crave more sweets, and gain more weight, and you have to contend with the added frustration that when you are dieting you are unable to lose weight. On top of all this, when you are irritable, everybody blames it on your diet—or the pills, if you're taking them.

What all this means is that you are five times more likely to go off a diet before your period than at any other time in your cycle—and to add insult to injury, your gynecologist tells you it's all in your head! Oh, if only men could have periods! The discomfort and the nuisance are just a *fraction* of the problem. If you are big-breasted, your breasts swell up to twice their size and weigh twice as much as usual. Your stomach pops out unmercifully for a week before. No matter how much water you drink, you excrete none of it. However little you eat, you seem to lose nothing. And as if all this weren't bad enough, diets even play funny tricks on the menstrual periods themselves, making irregular periods regular and vice-versa; and if you are on the birth-control pill, dieting can cause breakthrough bleeding. All this can drive you crazy!

Dieting can also change the time a woman ovulates, making the rhythm method of birth control unsafe. Weight loss, it seems, can enhance ovulation, increasing your fertility as you become more attractive—a dangerous combination.

I don't even accept the theory that premenstrual cravings are psychological. If you can get swollen breasts, fingers, and ankles before your period, you most certainly can get a swollen brain. When your brain swells, strange things happen that can affect taste, smell, and appetite; it is conceivable that the sensitivity of certain taste buds is heightened, and the one for sweets seems to be the most likely to be affected. Premenstrual irritability, too, is probably most easily explained as part of the "swollen brain syndrome." It is a well-known fact that irritability and various stages of confusion are common results of concussion when the brain is swollen from trauma; why shouldn't premenstrual swelling produce a similar effect?

How can you prevent weight gain before your period? If your weight is normal to begin with, all you need to do is go on the High Calorie Weight Loss Diet (see Chapter 13) one week before your period. You should also reduce your intake of salt, and avoid eating soups. If you are extremely uncomfortable, with sore breasts and backache, your gynecologist should give you a diuretic or water pill for those few days immediately preceding your period.

If you are overweight, just try to maintain your weight that week; that in itself will be a victory. Decrease the fruit in the Core Diet to one per day. For these few days substitute artificially sweetened beverages or foods for those containing sugar. The problem with using artificial sweeteners before your period is that your bitter taste buds are activated at that time just as much as your sweet taste buds, and you therefore tend to pick up the bitter taste of the artificial sweeteners. In order to counteract this, I'd recommend trying sweet and sour foods—sauerkraut, sweet-and-sour cabbage, fresh blueberries cooked with lemon and Sweet 'n Low; or coffee brewed with a cinnamon stick in it and artificially sweetened. If you can camouflage the bitter taste of the artificial sweetener (which the sour taste of lemon, for instance, does very well) you can make satisfactory low-calorie substitutions. A mild tranquilizer before your period will often help prevent you from overeating, because it makes you less edgy. However, I find an appetite depressant of little value during that week; all the work of an appetite suppressant is automatically undone by the premenstrual upheaval.

For the most part, the premenstrual week just has to be endured. Try and forget about it; work harder, play harder. Get more exercise. I think it's a good idea to avoid even keeping track of your periods, unless you are concerned about pregnancy. If you live your life around them, you will find yourself anticipating the symptoms (and perhaps making them worse) that ultimately cause you to eat more and go off your diet.

Once your menstrual period begins, the worst is over. Your weight goes down to its lowest; your appetite also drops to its lowest point in the entire month. If you've made it this far without doing too much violence to your diet, you're home free—for the next three weeks, that is!

Why *do* all these changes take place? You already know by now: it is the same estrogen-progesterone-ovarian-pituitary-fat water cycle that plagues women constantly. It is simply more pronounced immediately before and during menstruation—which only means that if you have a weight problem, you must train yourself to be even more vigilant with your diet than usual during this sensitive time.

CONCEPTION

While there are plenty of heavy women who get pregnant at the drop of a sperm, for those of you who are having trouble conceiving I would suggest that before you resort to fertility pills and all their attendant complications, you try a diet first. And if the desire to conceive is your reason for losing weight, then I would not take pills of any sort *except* thyroid, which in addition to its other virtues helps ovulation and increases fertility.

If you happen to get pregnant while you're on diet pills, don't panic. My male colleagues would love to pin birth defects on diet pills, but they haven't yet succeeded. A great many of my patients have inadvertently conceived while taking diet pills, and their babies have all been fine. Merely discontinue medication after you miss your first period until you find out whether you're pregnant or not.

PREGNANCY (or, Keeping Your Wife Pregnant Is Almost As Good As Keeping Her Fat)

The single most common cause of excessive weight gain in young females is pregnancy. When a patient returns to me after a year's absence and tips the scales fifty pounds heavier than when she left, I usually know without asking what has happened: She was given a license to stuff herself in the interest of childbearing. In fact, I've just read an article in a medical journal that said, in essence, "Gorge yourselves, girls, if you are pregnant!" This advice might be acceptable for a limited period of nine months at a time, and most women would certainly find it highly enjoyable, but the physical consequences of so much eating always, unfortunately, far outlast the pregnancy.

At no other time in a woman's life is she so dominated by the whims, idiosyncrasies, and personal attitudes of the (usually) male physician, and at no other time does she so blindly accept as gospel everything that he tells her. She unquestioningly obeys his every command, and all for the safety and protection of the unborn baby. My concern here, however, is the effect this devotion has on the physical condition of the mother, more especially her weight.

In one way nature has been kind to the obese pregnant woman; pregnancy increases her metabolism and makes it easier for her to lose weight. In the first trimester (three months), she needs about 150 extra calories per day; in her second and third trimester, she needs 350 extra calories per day to maintain her weight. Nature also kindly warns her of possible future trouble such as high blood pressure and diabetes. Pregnancy can cause a temporary hypertensive or diabetic state if she is overweight, which will revert to normal after she delivers. But this is an important warning.

For a brief period even obstetricians cooperated with the dieting pregnant woman, mostly because they felt that a certain disease called toxemia of pregnancy (whose symptoms were high blood pressure, edema, protein in the urine, and convulsions) was the result of gaining too much weight. Since this ailment endangered the mother's life as well as the child's, obstetricians encouraged, supported, and urged overweight pregnant patients to lose weight. This was the golden age of pregnancy. It was during these years, 1959–1966, that I was bearing children, and I used to tremble when I gained so much as a pound for fear of being scolded by my doctor. I well remember how, in 1959, when I was nine months pregnant, one of the staff doctors at the hospital, a pot-bellied surgeon, said to me, "Getting a little thick around the middle, aren't you, dear? A young girl like you shouldn't let herself go!"

In the past, a remarkable number of women kept their weight gain to a minimum during pregnancy and emerged from it with a healthy baby and a fighting chance to return to their prepregnancy weight. But the honeymoon was short-lived.

During the past seventeen years the attitudes of the medical establishment toward pregnancy and weight have done an about-face, and in a direction that has done the overweight female no good. Toxemia is no longer thought to be caused by excessive weight gain, and weight reduction during pregnancy, even if the mother is extremely obese, is not thought to be beneficial to the baby, although fat mothers face a greater risk of high blood pressure, prediabetic states, postpartum depression, and complications from Caesarean sections.

The present philosophy of childbearing encourages large weight gains (over 27 pounds, if you please) in the belief that heavier mothers produce bigger babies, and that the bigger the baby, the more intelligent. These conclusions were arrived at when it was found that pregnant women who were starved (as in concentration camps), or who were in the lowest socioeconomic levels and suffered from chronic nutritional deficiency, had smaller babies than normal. This is true, but bears no relation to normal dieting; however, male obstetricians are for the most part convinced that it does.

So now pregnancy has become a nine-month eating orgy. During the first three months of pregnancy, when there might be nausea and/or vomiting, you will be encouraged to eat six small carbohydrate meals per day, consisting mostly of crackers and soda, although there is medicine available to stop this particular side effect of pregnancy. If you are overweight, these little snacks are all you need to defeat any attempt at dieting. Clothes will be no problem, since the same designers who refuse to make beautiful clothes for obese women are determined to prove that you can be beautiful while pregnant. Modern maternity clothes can be elegant and of quality material, while they deliver the message, "It's fun to be fat when you're pregnant."

Is it any wonder that a female abandons whatever self-control she may have and indulges in her wildest food fantasies during this privileged time? Cravings for hot fudge sundaes and strawberries with whipped cream are immediately satisfied, without guilt. The husband, who may have been unwilling before to let his wife have so much as a snack between meals, becomes her willing accomplice, bringing home cartons of Chinese food and pastrami sandwiches. Oh, what joy to be eating for two! But remember, anything over a 19-pound weight gain stays with you after the fun is over.

Let's take a pragmatic look at some numbers. If a woman of normal weight gains 27 pounds during pregnancy (or an average of 1 pound per week for the last 20 weeks), she can expect that up to 19 pounds AND NO MORE will be lost when she delivers. With women of normal weight, a 27-pound gain is certainly not catastrophic because thin women have no trouble losing the excess pounds and do so quite efficiently as soon as they've had the baby. But add 27 pounds to a female who is already 20 to 30 pounds overweight and you have an enormous amount of weight to lose; it certainly doesn't help matters if you are told to eat more than usual to "keep up your strength." Acting on this advice, some of my patients have gained not 27 pounds, but 37, 47, 57 and even 67 pounds during pregnancy. These women may have healthy babies, but are they that much healthier than if they had kept their weight gain to a minimum? You'd have a hard time convincing me that they are. I wonder if anyone has even bothered to compile statistics on mothers-to-be who gain only 10 pounds on a high-protein, well-balanced reducing diet.

It is true that a 120-pound woman who gains 11 pounds during pregnancy has a smaller baby more prone to complications than a baby born to a 160-pound mother who gains 30 pounds. But what about a 160-pound mother who gains 11 pounds, and the 120-pound woman who gains 30 pounds? The point is, the cards are stacked against the fat female, and stacked in such a way that she can't argue the point even if she wanted to.

You do not lose weight very efficiently for about six months after pregnancy, and to complicate matters, you will probably become depressed. I don't mean the deep, intense depression that renders a woman unable to function, but the ordinary letdown that normally takes place after the "high" of pregnancy. This feeling can only be intensified by the realization that less than 40 percent of your 50-pound weight gain was baby. A slowing down of all metabolic systems also occurs, in reaction to the upsurge of those last three months, and it may take many more months for them to return to prepregnancy levels, if in fact they ever do; even at these levels they weren't so efficient.

I must formally protest the conclusions drawn from the available studies on dieting and pregnancy. In the first place, losing fat does not represent a nutritional deficiency provided that protein and vitamin levels are kept intact. Fat has no built-in properties that make you healthier, in spite of what mothers used to think. Nor can I understand the logic of comparing starvation to sensible dieting. Reduction of a mere 50 to 100 calories per day could result in a loss of 1 pound of fat per month, hardly starvation by any standards!

Nor is a female who consumes a lot of calories necessarily eating nutritionally. For the most part, she is eating carbohydrates, which represent empty calories. Some doctors fear that if a pregnant woman does burn too much fat, she will develop a state of ketosis (presence of partially burned fatty acids in the blood stream brought about in this case by an absence of available carbohydrates in the diet), which is not good for the baby. Unless you are dealing with a diabetic, a state of ketosis is very difficult to achieve in a female, pregnant or otherwise. One large salad or even a glass of skim milk per day would supply enough carbohydrate to take her out of ketosis, even if she were starving.

Well, what are we left with? With obese pregnant females who have increased chances of excessive bleeding, difficult deliveries, longer convalescences, more hemorrhoids, more diabetes, more high blood pressure, but enormous babies! Are we really prepared to sacrifice the mother's health for the hypothetical chance of a large genius? Of course we are! Your only problem after delivery is losing all the weight you gained over those nine months, and that isn't very important according to the present thinking of most obstetricians. I happen to think that it is very important, not only for the physical health of the mother, but for her mental and emotional well-being. It is a great deal more difficult to lose weight after delivery than during pregnancy.

But there is one bright spot: if you nurse your baby, you will use up even more calories than you do while pregnant. For between 600 and 1200 cc's of milk produced, you burn 700 to 800 calories of energy. This is your biggest bonus; you will never again burn calories so lavishly for doing virtually nothing.

Since medical fashion forbids me to preach diet during pregnancy (even though I think the conclusions drawn from the studies available are misleading), I urge you to diet wildly before you get pregnant! The only other alternative is to diet sensibly while you are nursing; either way, your efforts will be rewarded.

I have always had a suspicion that certain husbands believed that keeping their wives fat was better than keeping them pregnant because pregnancy is only temporary. Now possessive husbands can use pregnancy as the rationale for launching plump wives to new levels of fatness that will keep them house-bound and immobile forever.

HAIR

There are few nonfatal catastrophes quite so devastating to a female as losing her hair. The only comparable event in a man's life would be loss of potency. A woman's hair represents, symbolically, the very essence of her femininity and sexuality; no discussion of dieting would be complete, therefore, without discussing its effects on hair.

Does hair fall out because of dieting? Yes, sometimes; it also falls out following general surgery, abortion, shock, severe viral infection, from living and aging, and seasonally! In other words, it is as much in the nature of hair to fall out of your head as it is to stay in.

The life span of a hair follicle is about three years, and there is no evidence that sensible dieting will affect that. A scalp hair grows for about three years at the rate of about 1 centimeter per month, regresses for two weeks, rests for three or four months, and then is shed. Several weeks later, its life cycle starts again.

My patients' hair always falls out during certain times. Some of them correspond to periods of intense dieting, others do not. I suppose that if you severely depleted your body of protein (by total starvation), that might have some bearing on the health of your hair; or if dieting causes an iron-deficiency anemia, hair loss will also occur (and the anemia does not even have to be particularly severe). By and large, though, the health of the hair depends on the health of the hair follicle. If it is genetically programmed to put out good resilient hair for twenty years, it will do just that. At the twenty-first year, it will put out lousy hair, and all the diets in the world won't change it.

A patient of mine was worried that her hair was becoming fine, thin, and lusterless. When she consulted a noted dermatologist at Yale, he told her there was nothing wrong—her hair follicles were just putting out roller skates instead of Volkswagens. She was very annoyed. "He could have said, 'Fords instead of Cadillacs,' " she complained.

So being fat or thin won't make any difference in hair loss; however, if you are destined to have less hair, it will do far less damage to your appearance if you are thin than if you're overweight.

My hair is bad now—it died after a relatively minor virus. All the shots and vitamins with steroids and other hormones in the world won't restore it to health. All I can do is wait; most of the time, or almost all of the time up to a certain age, falling hair grows back; with luck, it's the same quality as it was when it departed. If your hair decides to leave, just bear with it; keep what hair you've got left clean, free of stress, and simply styled.

Sometimes when hair grows back after a sudden loss, the quality is poorer than before. But don't think that losing or gaining weight has more effect on the hair than any other body change; they're all unpredictable. If you start losing your hair, first make sure that there is no underlying physical disease. Then ride it out.

Normally, you can expect to lose up to one hundred hairs per day, and replace them at the same rate. If you lose more than this, you will notice that your hair is thinning. Eighty-five percent of your hair is in a growing phase at any one given time. Anything that disturbs the order of things, or changes growing, resting, or rejuvenation phases can alter hair balance.

I am told that bleaching and henna rinses restore some body to thinning hair; body waves are also useful. However, when my hair is falling out, I'm afraid to do anything to it. I try to put as little stress as possible on it.

Here are some of the suggestions I've received at various times from experts I've consulted to keep my hair from falling out:
1. You need thyroid (age 13).
2. You need iron (age 19).
3. Wash your hair frequently (age 20).
4. Cut it short (age 22).
5. Stop getting it teased (age 32).
6. Local application to the scalp of hormone solution, and shots (age 33).
7. Ultraviolet light (age 34).
8. Vitamins (age 36).
9. Zinc (age 38).
10. Stop dieting (ages 20–30–40).
11. Eat more protein (all ages).
12. Mild shampoo, nonalkaline or acid (age 39).
13. Don't color it (age 40).
14. Don't comb it wet (age 41). . . .

Get the idea?

My advice is to get one opinion from the best dermatologist you know. Then go home and relax.

SKIN

Aging in men is judged by two things: their ability to have sexual relations and their intelligence. Only when a man has become an impotent vegetable does anyone concede that he is over the hill. But beauty in a woman is judged primarily by the state of her skin.

Unfortunately, there is nothing worse for the skin than losing or gaining weight. This robs it of its natural elasticity, just as if you were blowing up and deflating a balloon. The skin of the face and neck is the most sensitive to weight loss, followed by the skin of the inner thighs, the upper arms, and the stomach. The faces of chronically overweight female patients age early; they develop jowls and crepey necks before 40. Fine lines appear later. Exactly when wrinkles will appear depends on the inherent elasticity of your skin. If you lose weight before you are thirty-five, and keep it off, your skin will usually be resilient enough (assuming you were not gigantically obese) to spring back into shape. Weight loss after 40 is much harder on the skin. Age and the amount of weight lost—not necessarily the speed of loss—are the criteria for determining whether you'll get wrinkles or not.

However, it is the speed of weight *gain* that actually ruptures the elastic fibers in the skin and causes stria, or stretch marks. These are most common in periods of high hormonal stimulation such as adolescence and pregnancy. They usually appear on the breasts, abdomen, hips, and occasionally the upper arms, and can be partially prevented by watching weight during those periods. Once you have acquired them they do not disappear, but they will fade from deep purple to white.

What can you do about wrinkles and sagging skin? I'm afraid creams are of limited value, drinking milk of no value. According to recent studies, zinc has some effect upon the elastic fibers; it's worth trying small doses of zinc, up to 30 milligrams per day. Exercising the flabby upper-arm muscles will help restore some elasticity; this is especially useful for the triceps muscle in the back of the arm, which normally gets no workout at all. I find the best exercise is to take a spray can, hold it over your shoulder in one hand, grasp the other end with the other hand, and pull. Then reverse hands. This puts tension on the triceps—but watch it, because it also aggravates bursitis. Do this about 10 times a day until you feel the pull in the back arm muscle. The idea is to develop the muscle so it gets larger and takes up some of the slack of the skin.

The small, fragile muscle on the inner side of the thigh (another bad saggy area) can be built up somewhat by lying on your back, pretending your legs are scissors, and crisscrossing them as far as you can 20 times a day.

Pendulous bellies, double chins, and jowls respond to no exercise except a walk or run to the nearest plastic surgeon's office. The best advice for women who are worried about their skin is either never to get fat in the first place, or (by far a second best) never to get thin.

WEIGHT DISTRIBUTION

Overweight women come in two basic shapes: pears and ice cream cones. Pears, the most common body types, have all their weight concentrated in the hips, legs, and thighs. Ice cream cones have most of their weight in their bosom, face, shoulders, and waist. These patterns of fat deposition are genetically determined; you will always find someone else in your family who is built like you. It would be nice if you could lose weight in the most unattractively fat parts of your body first, but women lose equally all over; having big hips doesn't mean you will lose inches there first and fastest. All the massaging and exercising in the world isn't going to make you lose more quickly from that one spot.

The theory behind massage is that by increasing the circulation to that part of the body, the fat will be carried away faster. Unfortunately, it doesn't happen that way. If you lose ¼" from your 42" hips, you will lose ¼" from your face. Your face will start to look thin, while your hips will look just the same.

However, the subcutaneous fat in a female is genetically programmed to be of a certain thickness, which means that if you lose too much from one area, your body will take fat from another available area. This is called redistribution; the process of moving the fat is called migration. I like to tell my patients that you lose from the head down, but redistribute from the feet up. You can never tell immediately after stopping a diet what the final form of your body will be, because this migration of fat takes about six weeks longer.

You can't see the migration and redistribution while you are still losing steadily, because layer after layer of fat is being eliminated too quickly for the process to be discernible. You will notice it only after you have slowed down or stopped losing altogether. Your face, arms, neck, and breasts usually appear to get thinner first, thinner than your lower extremities because up above there is less fat to lose. Then you will see your waist, hips, and stomach, and finally your legs decrease in size. Women who are 10 pounds away from their ideal weight will come into my office frantically demanding to know, "When will I lose these?" (pointing at their chubby legs).

"Last," I reply. And not just last, but last of all. If you get your weight down and hold it down for a few months, you will see an interesting change in your figure begin to take place up to six weeks after you're stopped dieting: those unsightly legs will get thinner and your face will become rounder. But you must tolerate a thin, drawn face while you are achieving a total slim look. Your face will take on a proper amount of subcutaneous fat after you've finished losing.

Many women let their faces be their guide when they diet, but you can see from what I have told you that that is not a good indicator. However, in the older female it might be well to stop short of pencil slimness, in order to preserve some additional subcutaneous fat in her face to prevent wrinkles.

HEADACHES

Many of my patients complain of headaches, especially if I ask them to starve for a day. The strange thing is that these same women have often gone without breakfast and lunch and have not eaten until supper, without getting a headache. But when *I* propose that they fast, a headache comes on about four o'clock in the afternoon, and instead of taking something for it, they use it as an excuse for eating. But since when is food a cure for headache? These particular headaches, I think, are the result of tension or psychosomatic factors, and I don't let my patients get away with taking the onset of a headache as a hunger signal.

Some headaches are probably caused by a release of histamine when you are dieting. These are pounding, migrainelike headaches; they are fairly common in the early stages of the protein-sparing and low-carbohydrate diets. They go away in time, just as the gnawing in your stomach disappears. In the meantime, I have used antihistamine with some success.

But the headaches that cause the most concern among my patients are those they attribute to their blood pressure. As I've mentioned before, many patients come to me with high blood pressure and when they get headaches after I've put them on diets, they usually fear that their pressure has risen. Headache is not commonly a feature of high blood pressure, unless the pressure is so high that the patient has to be put to bed or hospitalized, and in that case, many other more distressing symptoms than headache would also be present. Conversely, the blood pressure can go quite high indeed without a patient ever experiencing a headache. It is uncommon for my diet patients to have headaches due to elevated blood pressure.

When you have a headache on a diet, treat it just as you would a headache when you are not dieting: try symptomatic relief first, either with mild painkillers or tranquilizers; if the severity persists or is accompanied by nausea and/or vomiting, other causes must be found. However, headaches are common among dieting women, and are usually no cause for concern.

CARBOHYDRATE INTOLERANCE

Carbohydrates are the chief villains in the disease of overweight, because overweight women cannot tolerate them. By that, I mean two things: first, these women seem to convert carbohydrates into fat instead of energy; and second, carbohydrates make them feel both tired and depressed. This last is perhaps a result of the fact that carbohydrates make their blood sugar go up and down like a roller coaster. When you eat a piece of candy, your blood sugar rises rapidly during the first hour, which triggers an outpouring of insulin from the pancreas as the body attempts to lower the blood sugar. In many obese people this drop of blood sugar is quite rapid, and it is the abruptness of the plunge that makes some patients feel weak or tired. But they mistake the fatigue for a signal from the body for more sugar, so they ingest the sugar, and the whole cycle begins again. In graph form, you get a blood-sugar profile that looks like this:

blood sugar levels

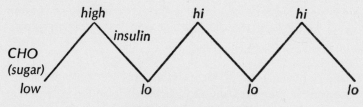

Roller-Coaster Syndrome

This roller-coaster pattern of blood sugar could be one reason for the low energy level of the overweight female, the immediate cause probably being the extremes of insulin levels as much as the blood sugar itself. It is also one of the reasons that obesity encourages adult-onset diabetes. Every time blood sugar rises, there is an outpouring of insulin. When this happens repeatedly over a lifetime, the cells secreting insulin can become exhausted, until one day there may be no more insulin to meet the continued sugar needs.

The other danger is the frightening and unexplained phenomenon that when overweight people eat a carbohydrate they cannot stop at one. It's no joke, like the potato chip commercial; it's true for bread, noodles, crackers, potatoes, and even more so for candy, cake, and pie. In its mysterious compulsiveness, it is very much like the alcoholic's inability to stop at one drink. Thin women do not have this problem, although they would like you to believe that it is their superior self-control that makes them stop at one or two pieces of candy. But most of the time, thin women are satisfied with one or two pieces of a sweet.

What exactly is operating here? What is this mechanism, apparently unique to the overweight female? I believe that the metabolic system of the overweight female has many defects which cannot, however, as yet be measured or quantitated. The vicious cycle of carbohydrate consumption leading not to satiety but only to more carbohydrate consumption cannot be explained by anything as simple as the "forbidden food syndrome," or on the basis of emotional problems. It cannot even be explained by the rapid rise and fall of blood sugar levels; the body has no natural food intelligence. All we can say is that it seems that, as with the alcoholic's relationship to alcohol, the foodaholic's intake of sugars and starches physically triggers off the insatiable craving for more sugar and starches, which is brought to a halt only with the physical discomfort resulting from sheer ingestion of huge quantities of food.

The most effective cure for the "roller-coaster syndrome" and the best corrective of the carbohydrate cycle and the carbohydrate intolerance of the overweight female is a high-protein diet. Protein is more satisfying in the long run than carbohydrates, and because it is filling, it burns more efficiently and gives her a much more even blood sugar level. This is far better for the whole metabolic system, for mood, and most important, for a sustained weight loss.

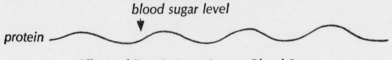

blood sugar level

protein

Effects of Protein Ingestion on Blood Sugar

QUITTING SMOKING AND GAINING WEIGHT

Another common metabolic complication that can create weight problems is the effect on the body of stopping smoking. The weight gain is usually not enormous, but ranges in the vicinity of 10 to 15 pounds, when the patient has been smoking over a pack a day. It is a phenomenon more observable in thinner women, but it is not primarily a result of eating more, even though nicotine is an appetite depressant. Since most thin women can normally burn off any extra food very easily, what is happening here is an actual slowing down of metabolic processes, one which, again, is relative to the individual, but is nevertheless frightening.

There are two ways to deal with it. I have often given minute doses of thyroid—5 micrograms, or ⅕ grain—in addition to putting the patient on the High Calorie Weight Loss Diet for about six weeks after she has quit smoking. Or else I do not use any thyroid, but put the patient on the Core Diet for three weeks, and the High Calorie Weight Loss Diet for three weeks, and insist that she drink four glasses of water per day to increase renal blood flow, and produce more diuresis. After six weeks, the patient's own metabolism seems to have adjusted to the lack of nicotine. Many patients also find that coffee or a food containing caffeine is particularly helpful after they've stopped smoking.

CONSTIPATION

Constipation means producing hard, dry stools which are sometimes painful to eliminate. Milk (a vastly overrated food for adults), a lack of fiber or cellulose in the diet, and dehydration are three of the main reasons why women become constipated during a diet. From infancy, when we are praised for our production of waste products, we place a high value upon substantial bowel movements. Overweight women tend to produce larger stools because they consume a high proportion of carbohydrates, and much carbohydrate residue is cellulose, which creates bulk.

It is an article of faith among many so-called nutritional and diet experts that daily bowel movements are essential to health and well-being, but certain diets in this book, particularly the Protein-Sparing Modified Fast, produce a small, skimpy stool only once or twice a week, because this diet contains no bulk. I do feel, however, that as a rule regular eliminations (not necessarily daily) help the patient to lose weight more efficiently. If constipation is a problem in the High Calorie Weight Loss Diet, I prescribe a simple lubricant such as milk of magnesia—three tablespoons nightly. In the Core Diet, where bulk is a problem, I recommend bran—two tablespoons in bouillon at four thirty in the afternoon (this also helps the patient to feel satisfyingly full). I usually find no other measures necessary. Incidentally, you may be cheered to learn that a good bowel movement can be worth two pounds of weight loss according to the scale.

8.

Dieting and Our Minds

Although I feel that it is mainly the intricacies of female physiology that play havoc with the weight-losing mechanisms of women dieters, I would not want to seem to underestimate the importance psychosocial factors play in women's dieting difficulties. However, I take issue with the popular notion that these are deep-seated emotional problems arising from the very roots of female development. Rather, they are a miscellaneous collection of misconceptions, taboos, and social propaganda, foisted upon us and perpetuated by those we love, those we respect, and those who ostensibly care for us. These psychological hangups can sometimes prove far more difficult to treat than physical problems.

COOKING

Most overweight women love to cook, talk about food, and read recipe books. Many have highly developed taste buds—they have, so to speak,

91

a talent for food. The trouble is, they always feel thay have to taste what they cook. If this is true of you, then I think it would be advisable not to do any ambitious cooking while you're dieting.

If cooking is a great pleasure to you, and you love having other people enjoy your food, I hate to deny you this means of relaxation. However, sooner or later the overweight woman must learn how to cook without falling victim to her own talent; it is unrealistic to expect her never to cook again. If necessary, she can stand at the stove with a plate of raw vegetables by her side, so that she can be nibbling on low-calorie snacks instead of the gourmet confection she's preparing for her family. She has to learn how to be around fattening food without overeating, but to begin with, she should always get somebody else to do the tasting for her.

One advantage of doing your own cooking is that you remain in control. If you have company, you can easily make yourself a salad with diet dressing and a plain hamburger for dinner while your guests enjoy the elaborate production you spent the afternoon making for them.

WOMEN, THEIR FAT, AND THEIR GENERAL HEALTH

Many overweight women have a very protective feeling about their fat; sometimes it seems that the more obese they are, the more lovingly solicitous they feel toward their flabbiness. I think it is related to the notion that fat is healthy and skinny is sick, an error passed on to us by our parents. And indeed it is true that before we controlled tuberculosis and the chronic infectious diseases, fat people, who could withstand the wasting effects of a long illness, were the ones who survived.

Today, however, I have yet to have one overweight patient in my practice lose any weight at all through any sickness of a nonfatal nature. Nevertheless, my patients get scared when they so much as lose their appetite from a mild case of flu. Instead of thinking, "Thank God, for once I'm not hungry," they stuff themselves still more. Women with colds eat crackers and chicken soup and Jell-O; women with diarrhea eat rice, bananas, and cooked cereal; women with gum trouble will eat mashed potatoes and gravy; women with bad gall bladders eat crackers and drink ginger ale. Women with no taste (from a cold) will eat anything fattening. And all of these women are supposed to be dieting! Every one of them could take advantage of her sickness to lose a little weight, and yet all jealously hang on to every ounce of fat, clinging to the excuse, "I need my strength." Most of my patients could go through cholera, the plague, and yellow fever in succession and still come out ahead in the weight derby.

As far as we know, fat keeps you warm but does not provide any special immunity to illness. Still, many of my patients will accusingly tell me, "I hadn't had a cold in years till I started this diet." Your diet *should* change when you get sick, but you should eat *less* and drink more (always low-calorie foods and liquids). Take advantage of the chance to give your digestive system a rest.

THE MARTYRED MOTHER

"Supermothers" always glory in the fact that they do everything for their children; fat supermothers are the greatest martyrs of all. They have so few other options open to them in life that the home becomes their hub, the kitchen their refuge, and the refrigerator their only source of pleasure. Is it any wonder they so love to bake bread, cookies, and cakes? They hide their ballooning bodies in the car, waiting for their kids to finish with the dentist or piano lessons so that they have an excuse to "grab" hamburgers instead of eating a sensible meal. Dieting is out of the question; how can den mothers, Brownie mothers, church mothers, think about their figures and their health when the children would get so upset if Mom didn't eat with them?

On the rare occasions when they do try to diet, the kids always seem to have gobbled up their diet food before they can get to it, or else they will ask, "How can I have meat while my poor children are eating a casserole?" Kids always adore a supermother, once they've got over the initial hurt of hearing everyone call her Fatso or Fatty behind her back. She's warm, loving, and always available, and many men with supermothers look for girls to marry who are just as plump and cozy as Mom. Afterwards, they can both go back home to super-Mom's for Thanksgiving and Christmas dinner.

Does it pay to be a martyred mother? Does it pay to sacrifice your body for the good of the children? Let's face it; whether or not it's worth it, the martyred mother *wants* to be the way she is—the children are simply an excuse to stay fat. It is not, after all, impossible to be both a supermother *and* a good dieter at the same time.

Here are some hints to remember:

1. Bring a can of water-packed tuna fish to those starch-loaded pot-luck suppers. Your child wants your presence, not your appetite.

2. Prepare ahead of time for nights when you're car-pooling or

watching Little League games. Have diet soda or a thermos of coffee in the car, and take along some hard-boiled eggs. If you can pack a lunch for the kids you can pack one for yourself.

3. Don't eat that piece of birthday cake. Nobody will remember or care.

4. Don't buy high-calorie foods *you* like "for the kids."

5. Don't *let* the children eat your diet stuff (or else get enough for all of you).

6. Remember, being a good mother has nothing to do with the food you put in your mouth.

MALE-FEMALE COMMUNICATION AND WEIGHT

From what I've observed, husbands and wives do not really talk to each other about weight; as a subject, it's more taboo than sex. They talk at each other about it, like parents and children; they get caught up in the mechanics of weight loss and weight gain, but never seem able to discuss the dynamics or reasons behind it. For instance, if a wife takes a forbidden piece of cake, a husband may ask, "Is that on your diet?" and the wife will probably snap back: "Mind your own business!" Or, "I can have a piece of cake if I want to!" Or even, "Of course it's on my diet!" After which both partners retreat into their corners in silence. Or, more characteristically, the conversation might go like this:

She: "I can't eat that. It's not on my diet!"

He: "I'm sick of that damn diet!" Or, "A little taste won't hurt you."

I would like diet discussions to begin along these lines:

She: "Why do you want me to eat when I'm struggling so hard to lose weight?" Or, "Do you know how bad it makes me feel when you try to get me to eat?"

My patients never say to their husbands, "Let's sit down and talk about my weight—why it's so hard for me to keep from overeating; how you feel about me being heavy; what you think your role is in helping me to be thin. How would our relationship change if I were thin? How would you feel about me? How do you *really* feel about my weight?" One wife told me that she and her husband argued constantly about her weight until one day she finally said to him, "I don't blame you for hating the way I look. You didn't deserve this. I don't understand why I allowed this to happen to my body, and I know just how disgusted you must feel." Her husband was so surprised that he almost fainted. He became transformed on the spot from a hostile, nagging tease into her most helpful and loving ally in a major diet program.

Communication between husband and wife on the deeper issues besides the simple do's and don'ts of dieting can be very beneficial in helping a woman lose weight, and in helping men to understand the special problems and obstacles women have in overcoming overweight.

SEX

Sex will never replace food in an overweight woman's life; that, alas, is a male fantasy. Heavy women will tell you that they are orgasmic, but on the whole I find them to be rather placid about the whole subject. I said "placid," not reticent; the feeling is generally that sex is somewhat overrated. However, this may well be the fault of the male partner. I hear frequent reports of little or no foreplay and perfunctory performance. Yet, strangely, few overweight women complain of feeling frustrated or bitter. Although they often find themselves physically unattractive, they do not seem to feel that their husbands do; when questioned about infrequency of intercourse they will usually blame business pressures, worry, or fatigue rather than their own unappealing bodies. However, they do report that they feel sexier after losing weight; when they are heavy, their desires seem to be buried in fat. They also feel that physical size alone prevents them from undertaking any very novel or ambitious form of sexual expression.

Being overweight does affect a woman's sex life, but not in the way or to the degree you might expect. The women who are the *least* overweight report the most trouble with their husbands. The women who are less than 20 pounds overweight seem to have more sexual problems with their husbands than women who are more than 20 pounds overweight.

Men who marry thin women generally expect them to stay that way, and get turned off if their wives become heavy, but men who marry somewhat obese women seem to stay relatively interested in their wives sexually no matter how much they weigh. This would be ideal, except that the husbands do not follow through with love, consideration, or attention; they seem to value their wives only as a means of satisfying physical needs. Some overweight women realize this and resent it—"All they want is a body"—but the majority see their husbands' enjoyment of them sexually as a stamp of approval of their size.

However, studies have shown that the husbands of obese women, when shown pictures of women of all body sizes, choose pictures of thin women as the most desirable.

Do fat women experience less sexual desire than other women? No, they start out with the same amount, but they quickly learn not to express their desires as openly, for fear of rejection. Fat single girls, no matter how beautiful, all report that finding partners they like can be a problem; many tell me that if they were thinner they would not be caught dead with some of the men they sleep with. Partners are always available, but the quality is poor.

Why, in an era when skinny is beautiful, would a man want to make love to a woman with hips that flop about the bed like giant cushions and a belly that sags unappetizingly to her dimpled knees? I wish I could honestly say that many men are mature enough to love a woman for her brains, her disposition, and her values, regardless of her body. But it's just not true. Some men "marry" their mothers (see "Super-Mom"). Eighty percent of my very obese patients report that their husbands' mothers were heavy, and let's face it, a boy's first love is his mother. Most men need mothers; some, on the other hand, don't care; a lay is a lay is a lay. Some men feel secure and comfortable knowing that few other men would want to make it with a fat woman; the fat wife is a faithful wife. This is one of the commonest motives behind a husband's insidious sabotage of his wife's diet.

The act of sex itself is a very minor problem, if it is a problem at all, when the wife is heavy; but sensuality, the feeling of being a loved and sensuous woman, is, as a rule, sadly lacking.

HYPNOSIS

Hypnosis, which can be defined as a state of increased suggestibility, has great potential in treating overweight women, by adding the extra ingredient that may help turn a wavering diet into a successful one. The only problem is that most adult female patients are difficult to hypnotize, which is related to their need, discussed in Chapter 4, "The Universal Overweight Syndrome," to be in complete control of the situation, although they deny this fact about themselves. This is unfortunate, since the role of the hypnotist in dieting is only to strengthen (or reinforce) the patient's own desire to diet. The ingredients for a successful hypnotic induction are: (1) a willing patient, (2) a patient who is not afraid, (3) a patient who trusts the hypnotist, and (4) a patient who is able to concentrate. This sounds simple enough, but my first two bouts with hypnotizing overweight females proved more difficult than I expected.

Dr. Herbert Spiegel, a well-known New York psychiatrist and authority on hypnosis, had instructed me that even if the hypnosis wasn't working, I should act as though any distractions were part of the process. During my first attempt, someone began banging at the door to my office, so I solemnly said to my patient, "While you are in this trance you will be aware of someone banging at the door, but that will not disturb your relaxation." My patient opened one eye and said, "Like hell it won't!" That was the end of that.

My next attempt was with a woman who had expressed interest in hypnosis for a long time; she was certainly willing. As I launched into my induction, I noticed her mouth grimacing. I tried to ignore it, but she kept on doing it. At last I said, "Sometimes during hypnosis you may find yourself making faces, but that's quite all right." At that she opened both eyes and said, "My stomach hurts!"

In spite of my two false starts, I have used hypnosis successfully. Not as often as I would like, though, because it is time-consuming and many patients are still afraid of it. This fear is not as unrealistic as it may seem; unquestionably, hypnosis in inexperienced or unethical hands can present problems. Anyone can practice hypnosis; just make sure, by checking with your local medical society, that you are going to a legitimate practitioner.

Many women fear hypnosis, but some of their fears have little basis in fact. It will not drive you crazy; you will wake up afterwards; you won't simply replace one bad habit with another. However, male seduction of the female under hypnosis has been known to happen, and the argument that the patient doesn't do anything under hypnosis that she wouldn't do in a waking state is not much help when a therapist uses hypnosis plus his privileged position for unethical purposes.

Teenage girls are ideal candidates for hypnosis as an aid to dieting. They have not yet developed most of the hangups of their elders, and they are willing, suggestible, and usually trusting. The problem, if any, is the mother. Let me tell you about two cases:

A very obese 14-year-old came to see me with her mother. It was obvious from the outset that the child wasn't really ready to follow any diet; all she wanted to know was when could she have ice cream, peanut butter, and mayonnaise. My immediate thought was, why not hypnotize her? I carefully explained the process of hypnosis to both the mother and the child, telling them that it was a state of increased suggestibility, during which the hypnotist would make certain suggestions relating to diet.

After consulting with her pediatrician, the mother brought the child back the following week. She proved to be a model patient. When she was in the trance state, I suggested that it would be nice for her to be thinner, and to help her accomplish this, I would give her a diet to follow. I kept emphasizing that her desire to be thin would be greater than her desire to eat fattening things. "You have only one body," I intoned, "you owe it to your body not to eat food that will keep you fat," and so on. (A fundamental rule of hypnosis is never to suggest to the patient directly that she hates sweets, or that chocolate cake will make her sick; such a manifest lie won't fool even the unconscious.) As the mother watched the hypnotic process, she seemed interested and to understand what was going on.

After the child was brought out of the trance, her mother took her to the snack bar downstairs. The child ordered a roast beef sandwich, a pickle, and a diet soda. As soon as they got home, her mother called me to report her amazement; normally the girl would have demanded a grilled-cheese sandwich, potato chips, and a milkshake. I could tell the mother was slightly worried, though, as if wondering, "Why should my child do that?" A week passed uneventfully. I could scarcely wait to see her again—my first success! On the day of her appointment, no patient appeared. I usually don't call patients, but I had to find out what had happened.

It turned out that during the second week of the diet, when everything was going well, I went away for a few days to a meeting, which coincided with the mother's birthday (a circumstance I had not foreseen). She asked the child to take a "sliver" of cake, which she refused. The mother insisted and the girl began to cry, but still refused. This behavior was so untypical that the mother panicked and called their family doctor, who said, "Force her to eat the cake!" The mother was so frightened of the hypnosis that she ended by sabotaging the whole program (with the help of the doctor). Anxiety created by the conflict between mother and child dissipated the effect of the post-hypnotic suggestion—not an uncommon happening.

The second story is somewhat similar. In this case, the girl was 18 years old and weighed 250 pounds. She had tried every system of losing weight, but could never manage to stay on a diet longer than a few days. In desperation, I used hypnosis. She was a perfect subject, willing and suggestible. I suggested to her simply that she follow her diet and stick to it. Two weeks later, she was back with a 17-pound weight loss. I put her in a trance again, and suggested that she continue to follow her diet. Next time, she came back with a weight loss of seven pounds. As I was preparing to induce trance once again, she said, "Stop! I don't want to be hypnotized! I'll do it on my own." I stared at her in disbelief, then it dawned on me. "Who have you been talking to?" I asked.

She paused. "My mother," she said reluctantly.

"And what did your mother say?" I asked.

"Well, when I left your office the last time, I went home and began throwing up."

"What was the matter?"

"I think I had a stomach virus or something."

"But your mother thought I had hypnotized you to throw up, didn't she?"

"Uh-huh," was the sheepish answer. Needless to say, the girl's weight-loss program deteriorated rapidly after that.

In both cases the mothers' fear and misunderstanding of the process of hypnosis had destroyed its potentially great therapeutic value. This makes me angry because hypnosis is both a useful tool and a harmless one, if used correctly. I still feel that for a good subject, hypnosis is the best way to get psyched up *and* stay psyched up for dieting.

GROUP THERAPY

During my psychiatric residency, we used to say that anybody who did group work with adolescents had to be a masochist. I've come to think that the same must be true of anyone who works with groups of overweight women. Consciousness raising and therapeutic group work are an even more popular adjunct than hypnosis, but in my experience they are far less successful. Overweight women are highly reluctant to be honest about and discuss their feelings, and tend to be passive and noncommunicative, except in areas of weight, food, recipes, and other "how to's" associated with dieting.

In the groups I've led, when I tried to steer the conversation from dieting to feelings of anger, depression, and self-contempt, the overweight women retreated behind a blank exterior. They were unable to show anger at other group members even if they deserved it; neither could they disagree with the group leader. My feeling is that the overweight female responds best to a one-to-one relationship where you can challenge her, refute her without embarrassing her, and compel her to come to grips with herself, her own tricks and evasions. The group is usually too passive to confront her in a useful way.

9.

Men Who Help and Men
Who Hinder Dieting Women

1. DOCTORS

The two most important men in the life of the universal overweight are
her doctor and her husband. Both are in a strong position to help her;
most of the time they do not. The first male doctor the female sees is
usually her gynecologist. He most often is not interested in doing weight
work, and to his credit, he tells his patient so frankly. But he usually *can*
be talked into giving her a month's supply of diet pills or diuretics. When
she appears 30 days later with no weight lost, asking for more pills, he
backs off and says, "That's it!" At last gynecologists have begun sending
their patients to diet clinics, which is great, but I wish they did it to begin
with, *before* they wasted the effectiveness of diet medication.

The internist is the physician who sees more women on diets
than anyone else, and he is often their worst enemy. "What do you

mean?" demanded one of my internist friends, when I said this to him. "I have three women on diets who are all doing beautifully."

"In spite of you, not because of you," I retorted.

Most internists either consider it beneath their dignity to treat dieters, or else feel that dieting counsel is not the province of the physician at all. When I was at a diabetes meeting, I heard one of the major figures in the world of diet research say that he doubted whether weight control should ever be in the hands of doctors. My feeling about that statement is that if psychotherapy is still in the hands of psychiatrists and psychotherapists, and deliveries are in the hands of obstetricians, and diagnosis in the hands of the internist (who may soon be replaced by a computer), then dieting can be in the hands of the doctor.

Let's look at some of the different physicians who directly or indirectly concern themselves with weight control.

The Purist
The purist doesn't believe in pills of any type; he pushes willpower. Personally, he feels that all fat people are the same; anyone can lose weight, with the application of a little willpower. The only help the purist can muster up is a diet sheet and a short pep talk on how easy it is to lose weight (after all, *he* did it). Purists are a hypocritical lot—many of them have wives who can lose weight effectively only by taking diet pills, but the purists turn their heads the other way and ignore this when dealing with patients. They diagnose only by their laboratory tests; if the lab tests are normal, then the patient must be.

The fact is, they don't want their offices cluttered up with overweight people. At a meeting I heard one internist say, with rare rationality, that the newer diet pills might have some use in short-term dieting. The purists were all on their feet in an instant, demanding, "Do we understand that you believe in diet pills?" Unfortunately, the man on the podium backed off and mumbled weakly, "I didn't say that!" If you can lose weight with a purist, you could lose with anyone. (And it must be said that many people, if they are ready, *can lose weight with anyone.)*

Most male physicians will sabotage overweight females, using any flimsy excuse; they do it, of course, in the name of helping them. They tell heavy patients to eat ice cream for sore throats, or to forget about their diet when they have the flu or an upset stomach, or to eat after surgery to keep their strength up. All of this advice was given to women who weighed in excess of 200 pounds.

You will in the course of a major diet get two or three illnesses of various types, from flu to strep throat to gastrointestinal virus; you may in the course of a major diet have elective surgery. There is absolutely *no reason* in any of these cases to stop your diet. Yet internists constantly tell my patients, "Forget about your diet until you get better." This is *precisely* what the overweight female patient wants to hear. Ice cubes are fine for strep throat; plain codeine, not in syrup, is fine for a cough; and potassium iodides in water do a good job in liquifying mucus in the lungs; tea, diet soda, water, or bouillon are great for supplying liquids, and vitamin C tablets and tomato juice (not high-calorie orange juice or "fruit juice") are perfect for colds. An overweight person does not have to "feed" an illness. Her weakness stems from the *illness,* not from *lack of food.* Rest, low-calorie fluids, vitamins, and perhaps small amounts of boiled chicken or poached eggs for protein are all you need when you are sick; in fact, you need fewer calories than usual. Use illness as an occasion to rest from food, not as a reason for pushing your sluggish digestive system to the limit. But if you do try to continue your diet, you will probably meet with resistance from your doctor; male physicians always give sick overweight females a license to overeat, and may become annoyed if you refuse.

The Reformed Purist
I must include this category because male doctors can be like chameleons. The reform purists once pushed diet pills, water pills, thyroid, and God knows what else, but when they saw it wasn't politically expedient, they stopped and went into reverse. Like reformed smokers, the reformed purist—feeling guilty about his former practices—preaches just as hotly against the thing he once adamantly supported. I don't mind a doctor admitting an honest mistake and changing his mind when new facts emerge, but I do resent these opportunists who become evangelists for the other side just because it is now fashionable.

Research Buff
These are the worst. They write scholarly books, but never see patients. They either do basic animal experiments or tabulate clinical studies, and they never have the slightest feeling for the overweight patients themselves. These people may have the answers for us *tomorrow,* but to pass themselves off today as all-around experts, when the closest they ever get to patients is a hospital visit to a friend, is annoying.

Diet Doctors

On the other end of the scale are the diet doctors, the M.D.'s and D.O.'s (Doctor of Osteopathy), who limit their practice to weight loss. They are almost always huge financial successes, which gives some indication of the great need for them. But they have prospered only because the internists have turned their backs on the problem of obesity, or tried to delegate it to diet clubs. Internists rarely refer their patients to diet doctors because that would mean giving up control; they much prefer to recommend some nonmedical discipline like a diet club, which doesn't threaten their authority over the patient.

No special advanced medical training is needed to become a "diet doctor." You may start as an internist, and I have also heard of gynecologists and surgeons reviving sagging practices by treating weight patients, but diet doctors are usually general practitioners who for some reason have decided to limit their practice to weight control. I wish I could say that all diet doctors know what they are doing, but unfortunately they run the gamut from good to terrible.

The bad diet doctors have brought the reputations of the good ones down with them. Because it is an easy way to build a fast practice with a minimum of night work, rare emergencies and little or no hospital work, diet is a haven for shrewd, opportunistic, greedy, and poorly qualified men. It is also a refuge where doctors who are tired of the rat race of medicine can settle back and make a good income. It is interesting to see how women pick their internists and gynecologists with such care, but will go blindly to any diet doctor who hangs out a shingle. Everyone has heard horror stories about diet doctors—the arrays of rainbow pills and patients getting very sick—and, as stated earlier, one unscrupulous group of doctors even used diet pills with digitalis in them, an inexcusable practice. A good diet doctor can be a savior, but a bad one can cause you severe discomfort and even illness.

One diet doctor I knew of who practiced in a nearby town was extremely busy. Several of his patients who had left him came to me and reported with some amusement that he did monthly pelvic exams, which took a long time. They wondered what he was checking for. (Weight loss in the vagina?) I asked them if they had called the medical society and reported his practices; only a few had bothered. It took three years of complaints before this doctor was asked to leave town. The point is that patients will tolerate long waits, inadequate treatment, and at times even unethical behavior just to see a diet doctor. Why? Because the need is great, the patients are desperate, and the purists are pompous and unsympathetic.

How does one find a good diet doctor? First, you can call your local medical society and ask them to recommend one. If the medical society can't help, then you will probably go to someone you hear of from a friend (using what we physicians call "word-of-mouth referral"). I would tell you to ask your internist, except that most of them are so uptight from fear of either losing control of you or losing you altogether as a patient that they will try to send you to a "safe" diet group. But try asking him anyway; you may be lucky and get an unprejudiced answer.

If you do find a diet doctor, make a checklist for the following:

1. Look at the office: is it a legitimate doctor's office, or does it look like a fly-by-night operation?

2. Find out what medical school your doctor graduated from, and where he took an internship and residency.

3. Look out for gimmicks, such as products for sale on the side. Reputable doctors don't turn their offices into a marketplace or a pharmacy.

4. Does he give you a physical exam? It doesn't have to take an hour, but it should be at least 20 minutes, and should include blood pressure, pulse, heart check, lungs, thyroid exam, abdominal exam, neurological exam. Some physicians do a rectal and proctoscopic exam, which is good but not necessary, since a diet doctor is usually not a primary-care physician (a family doctor). A pelvic exam is rarely necessary before starting a diet.

If the patient has not seen a primary-care physician in over a year, a breast exam is a good idea. I've found too many lumps not to do breast exams routinely.

5. Does the doctor or nurse take a good history, especially about food intolerances, diet history, family history? If the nurse takes it, does the doctor discuss it with you afterwards?

6. Is blood work ordered? EKG if you are over 35? If not, why not?

7. Do you get a chance to see the doctor personally to discuss his diet recommendations with him, or do you listen to a recording, or get interviewed only by a dietician or a nurse?

8. Do you know what pills you are taking? You should know their names, actions, side effects, and what you are taking them for. There is rarely a legitimate need to take more than two or three pills a day (not counting a vitamin pill).

9. Do you get shots? What kind? Why are you getting them?

There is little need now to give medication by injection except on the H.C.G. diet, which may be useless anyway (see Chapter 10).

10. Do you come back to see the doctor on a regular two to three-week basis? A diet is a *program*, and part of the program is regular visits, preferably every two weeks; it is difficult to maintain continuity on less frequent visits. When you return for subsequent visits, do you see the doctor? The same doctor? Do you have time to tell him your problems or ask him questions? If you don't, you are not in the right place. The one thing you need most from a diet doctor is concern, in the form of attention and expertise, *not* just pills.

Too many patients go to a diet doctor *only* wanting pills. If you go to a diet doctor for that reason alone, you are missing the whole point. If pills were all that mattered, the women who got a month's supply from their gynecologists would have lost weight. The key is pills *plus* professional help.

11. Are you free to call up and talk to your doctor if you have a problem? That is essential.

12. Are his fees in line with his services or with those of other doctors you see? Is he willing to talk to your primary-care physician if you want him to? You'd be amazed at the number of intelligent women who guiltily go sneaking to a diet doctor because they are afraid their regular doctor will get angry with them if he finds out.

13. Does your doctor tailor your diet to your specific needs, or does he give the same diet to everyone?

14. Does he know you? Do you feel that he likes you? Make sure you are not just a number to him.

Like little girls with little curls in the middle of their foreheads, diet doctors can be good, or they can be horrid. The doctor who puts patient volume and personal gain ahead of the welfare of the patient will serve you only after serving himself— beware of him. All doctors are entitled to fees for service; just make sure you are *getting* service.

Currently a great many other specialists have started jumping on the diet bandwagon—psychologists, behaviorists, hypnotists—and there is a place for every one of them. However, I see no need for the dentist in this area. First, because I think wiring jaws is one of the most inhuman tortures invented since the Inquisition; and furthermore, for intelligent men to be convinced that it has any merit as a means to dieting is unthinkable. Besides the annoyance and the risks, the whole premise is ridiculous. My dentist friends who got talked or cajoled into wiring jaws are now disgusted with the whole procedure and wonder why they ever did it. Almost all of these patients cut their own wires in panic, hunger, or anger, after a short, dramatic weight loss, and gained all their weight back even more quickly.

Surgeons

In a separate category are surgeons, who are in a unique position to help—or harm—fat women. At least 50 percent of my patients sport ugly surgical scars on their abdomens, the usual culprit being the gall bladder. Gall bladder disease, due either to stones or to chronic infection, is known to be more frequent in females who are "fair, fat, and forty." The usual treatment is removal of the affected organ, which is why I see so many surgeons' signatures glaringly carved across my patients' abdomens.

This infuriates me; first, because the surgeon who deals with gall bladder disease has both a right and a responsibility to discuss diet with his patient, both pre- and postoperatively. Preoperatively it is easy to reduce a patient because many fattening foods make her physically sick; a simple, low-fat, low-carbohydrate diet is therefore both therapeutic and desirable—and her losing fat can only make surgery easier for the doctor, too. After the operation, the surgeon has a unique opportunity to discuss eating habits with his patient; it would be extremely easy for him to help her lose weight in the postoperative period while she is still under his care. It's too bad that so many surgeons suffer from tunnel vision, focusing only on the organ to be removed and neglecting the whole patient.

Another annoyance about abdominal surgery is having to see, in a society where such emphasis is placed on beauty, bellies scarred like relief maps of the United States. I've also seen gall bladder scars from nonemergency surgery in which the surgeon could have taken out the tonsils and hemorrhoids through the same incision. One of my surgeon friends makes the excuse that it's important to have "good exposure," meaning to be able to see the organ clearly. Nevertheless, I sometimes feel that if some of the guys weren't racing the clock to see how fast they can take out an organ, they wouldn't be making incisions as long as the equator.

Gynecologists are starting to see the light; they're trying to keep their abdominal incisions within the pubic area, and many are doing more surgery through the vagina. But it seems to me that general surgeons could do much better jobs when they perform elective surgery on an overweight female patient if they encouraged her to lose weight beforehand, and had a little more respect for the appearance of her body afterwards, unappetizing as it may seem to them.

2. HUSBANDS

Husbands are extremely important to dieting wives, but, as we've seen, are often not as helpful as they should be. The husbands who are going to resist strongly their wives' getting heavy will make their objections known with the first 20 pounds, and those wives, if they value the stability of their marriage, will immediately lose weight. Luckily, husbands like this have wives who are vain to begin with, and there is usually no further problem.

More complicated are the men who marry chubby women and let them become grossly obese with never a word—all in the name of love. "Oh, my husband never says anything to me," one 250-pound woman told me. "He is afraid he'll hurt my feelings." I had long wondered why so many men never say a word to their wives as they get heavier and heavier, and I found the answer when I saw some of these women lose weight: many husbands are threatened by an attractive wife. As one woman put it, "my husband told me when we were dating, 'I'm going to marry you and make you so fat that no other man will ever look at you.'" And that's just what he did.

Many men, without verbalizing it, feel this way. It is comfortable to have a fat wife. She is usually a good cook, a fine mother, and an efficient housekeeper. She makes few demands on his lovemaking talents and energies because she doubts her own sexuality. She rarely buys clothes because it is too embarrassing, and is content to stay in the house for weeks at a time because she's ashamed to have people see her. She rarely goes into the job market, so she offers him little competition, and when she does get a job it is expected to be some menial kind of work that will not interfere with her home duties. If you think I'm being too harsh on men, you'd think differently after you'd heard the same pattern repeated over and over, as I have. Men still feel that they own their wives, and overweight women are the most "owned" of all. Many of my patients are afraid to tell their husbands that they're coming to see me; when they finally do, their husbands are likely to say, "You can go as long as you lose weight," or, "I can't understand why you can't do it yourself. All I do is cut out desserts."

When such a woman starts to lose weight, her husband might say: "Have a taste of this chocolate cake—just a little bit won't hurt you."

"I can't stand you being so cranky. Get off that damn diet."

"*Do* you look better? You looked okay before."

Or, when they go on vacation: "Don't spoil *my* vacation by dieting."

"I'm paying for this food. I expect you to eat it."

In the final stages of dieting: "You're getting to be a fanatic. Your face is too thin."

I have seen many marriages break up because the wives decided to take themselves in hand. They begin to see themselves as desirable human beings, and they resent the fat trap their husbands have locked them in. One wife told me her husband would always bring home boxes of candy when she was dieting, pretending to be courting her. Finally, one day she threw them all into the trash compactor, unopened. He didn't try that again.

Overweight women *must* assert themselves and let their husbands know that they still have the right to control their own bodies. If both parties accept this, there is no difficulty; but mutual respect and help for the dieting wife is imperative for success.

10.

Winning the Losing Battle (or, My Life in Dieting)

Becoming a diet doctor was a natural step for me, although I never planned it. I had completed a year and a half of a psychiatry residency before deciding that psychiatry wasn't the specialty for me. I was interested in *actively* treating patients, but responsibilities for a young family made me think twice about pursuing a career in internal medicine or general practice. For a girl who once had won an award as the most valuable female medical school graduate in Philadelphia, I was not accomplishing much. Finally I went to work part-time for a busy general practitioner.

One of the first patients I saw was a young female, 22 years old, 5'10", who weighed about 200 pounds. She had come in for a physical exam. She also wanted to lose weight. I put her on the High Calorie Weight Loss Diet (see Chapter 13). She was a *perfect patient*—she was young, well-motivated, and it was her first diet.

After about four months she had lost 40 pounds, and something very peculiar was beginning to happen to me: my side of the appointment book was filling up with people who wanted to diet. "What an ideal specialty for me," I began to think. It combined psychiatry and general medicine with a subject that was of special interest to me: overweight. It also lent itself to a part-time practice—I could work and still fulfill my family obligations. So my diet practice was born!

Being a diet doctor has been both frustrating and rewarding. Had I not been used to the slow pace of psychiatry, I would have become discouraged. But the universal overweight personality type, with all its stubborn resistances, is only a part of the total person—the diseased part. It has been a challenge to try to bring the sick part of the personality toward health, while improving the appearance, function, and productivity of the rest.

Personal experience taught me what I know about diets. Some diet techniques I discarded as ridiculous; others I kept and incorporated into my standard regime. What follows is a summary of recent diet theory and practice, based on my own experience with most of the major diets.

—20 BMR DIET

At 16, when I found out that my basal metabolism rate was low, I put myself on a crash diet I had seen in a magazine. You could eat up to 12 oranges a day and drink a soup made from stewed tomatoes, cabbage, and celery, prepared without salt. I lost 10 pounds in two weeks and discovered the following:

1. A diet without protein makes you tired.
2. A diet without protein makes you look haggard.
3. If a diet is monotonous enough, you end up eating less than you are allowed. I could manage only six oranges a day by the end of the two weeks.
4. You lose about 10 pounds in two weeks on a crash diet.
5. Two weeks is the maximum time one can tolerate a crash diet.

LIQUID DIET SUPPLEMENT

I went on my next diet in medical school. After I had my first child, my weight had crept up a little. I needed a diet that was relatively cheap and easy, and a liquid supplement had just come out on the market. I drank 4 cans of it a day for 2 weeks and discovered the following:

1. I lost the usual 10 pounds.
2. The high carbohydrate content in this purely liquid diet stimulated my appetite, and I was constantly hungry.
3. The milk content of the supplement alternately caused bloating, abdominal pain, and gas.
4. Diet supplements such as bars, wafers, and liquids can be valuable if you want a fast source of balanced nutrition and have no other way to get it. But I feel that the carbohydrate content is too high for most overweight females to lose weight efficiently on supplements. I later found that the Liquid Supplement Diet did not satisfy most women; it stimulated their appetite, so that they ate regular food in addition to the supplement, and gained weight instead of losing it. And I discovered that if anything tastes too good (like a chocolate milkshake) fat women will eat or drink too much of it, and will consequently not lose weight.

PILLS AND RASHES

After the birth of my second child, I had my first diet experience using pills. (I had previously taken diet pills to stay awake at night to study, but never to diet.) I had gained about 30 pounds, so I went to the internist who was then the fasionable diet doctor. He gave me both diet pills and diuretics. I lost 15 pounds in one month on his diet, and then I stopped losing altogether. He did not change the diet and we seemed to have reached an impasse. I did notice, though, that one of the pills was giving me a terrible stomachache, and I also began to see big blisters on my feet. I went to a dermatologist who biopsied the blisters and said they were caused by an allergic reaction to the diuretic. I stopped that diet and learned the following:

1. I had no side effects from taking the diet pills and had no trouble giving them up—and they were valuable in controlling my appetite, at least for the first month.
2. Even though your doctor is responsible and cautious, you may have a problem with medication—but it could be an individual problem. Whenever you put a pill in your mouth, you can never be absolutely certain how you will react to it. That doesn't mean that there is something wrong with the pill or with you; it means that there are

inherent problems in taking any kind of medication, so if you are using any pills to diet, you should be supervised!

3. Diuretics can cause even more problems than diet pills.

4. Diets may need to be modified as women lose weight; the diet that helped you to lose the first 20 pounds may not be as effective for the next 20.

H.C.G. DIET

Several years later I began hearing of a magical diet from Italy which used Human Chorionic Gonadotropin (hormone from the urine of a pregnant woman) for weight loss. It was the rage in California, and also was widely followed in Florida. You received a daily injection of this hormone in your buttock which was supposed to: kill your appetite, mobilize your fat stores from your fattest spots first, and make you look terrific while you were losing weight. The only catch was that the injection had to be supplemented by a 500-calorie, absolutely *fat-free* diet. You couldn't even use cream on your face, and if you were handling something greasy, you had to wear gloves.

The program consisted of six weeks of daily injections (six out of seven days—Sunday was a day of rest), followed by six weeks of maintenance, followed by another six weeks of daily injections, and so on until you lost as much weight as you wanted to.

The opponents of this diet argued that anyone could lose weight on 500 calories a day; its adherents claimed that the injection killed your appetite completely, made you feel euphoric, and enabled you to follow the 500-calorie diet. The first six weeks I lost 24 pounds; it was so exciting I couldn't bear to give up the weight loss for the second six weeks and go on maintenance, so by fasting one day a week and following my Core Diet, I lost 2 pounds a week for another six weeks The third six weeks I went back for shots. During the next cycle I followed everything exactly as I had the first six weeks, and lost only 12 pounds. From this diet I learned the following:

1. H.C.G. did not seem to hurt me; I saw no harmful side effects in myself or others. It could be safely given to high-risk patients unless they were in such bad shape they couldn't stand the needle.

2. H.C.G. might have a slight appetite-killing effect, because it may supply a touch of nausea (it is, after all, the hormone of early pregnancy).

3. H.C.G. actually *hampered* weight loss by encouraging fluid retention, or perhaps even by stimulating fat production—on a

500-calorie diet you could lose weight faster without the shot. I saw no selective weight loss from the fattest spots first; the pattern of weight loss was exactly the same as on any other diet.

4. H.C.G. has no particular merit in dieting except as a gimmick; I should also mention that a few girls who were on H.C.G. and were having infertility problems got pregnant soon afterwards and gave birth to healthy, normal babies. So perhaps it has some value for the dieting female who is trying to get pregnant.

5. You do look great on the H.C.G. diet (younger, and with healthier skin and hair).

THE REVOLUTIONARY DR. ATKINS

I first met Dr. Robert Atkins, of low-carbohydrate, "unlimited calories" fame, at a Columbia University conference on nutrition. Dr. Atkins' "Diet Revolution" claimed to achieve dramatic weight loss by allowing dieters to eat unlimited amounts of high-protein and high-fat food as long as the carbohydrate intake was drastically cut down. "Stop counting calories forever" was the cry. The theory behind this permissiveness was that the food that was allowed created extremely high levels of fatty acids in the blood, called ketones, which acted as a natural appetite depressant, so that in practice the caloric intake would be restricted as well, although the patient was able to feel that he was eating as he had never been able to—for instance, steak with Béarnaise sauce, lobsters swimming in butter—and still lose weight.

When I first saw this revolutionary, he was arguing with one of the "purists," who finally said to him, "Sit down, Dr. Atkins. The body does not spill out excess calories, and that's that!" I couldn't wait to go over to Atkins and get in my two licks. "I think your diet is great for men and for maintenance," I told him, "but it doesn't work on fat women."

"That's because they are not following it," he snapped back at me crossly.

"They tell me they are," I replied lamely.

"Then they're lying," he said, as he walked off.

I don't think so, Dr. Atkins. I think you are a brilliant doctor who had one of the few original ideas in dieting, but for the chronically overweight female, it is a bust. It's a great diet for men and thin women, but after the initial 5- or 10-pound water loss, overweight females eat right up to their calorie requirement. OVERWEIGHT FEMALES CAN NEVER EAT ALL THEY WANT, and since they resist ketosis (it is difficult for them to get that high a ketone level in their blood), they rarely stop being hungry before they have eaten too much. That's what happened to me on your diet, Dr. Atkins, and I didn't cheat and I didn't lose weight.

Later I had the opportunity to spend a day in Dr. Atkins' office. His caseload reminded me of my own—upper-middle-class and predominantly female. He, too, had women patients who didn't lose weight on his original diet, and he was forced to decrease the amount of food they could eat. He explained to me that these women were unusual in having metabolic abnormalities—but I think all fat women have metabolic abnormalities.

Dr. Atkins is a true pioneer in dieting, and the stuffy AMA stand, which permits only so-called sensible, "balanced" dieting, typifies the narrow-mindedness of the purists and their resistance to new ideas. But I don't think the doctor really understands fat women. Although, when I was ready to go home, he asked me, "How did you enjoy your day in my office?"

"It was fine," I said. "It reminded me of my own office, except that this city makes me nervous."

"That's because you eat too many carbohydrates," he quipped, as he ushered me to the door.

BLESS DR. BLACKBURN

I met Dr. George Blackburn at a conference on Nutrition and Adolescence. As usual, the major speakers were men, all well-known and full of the traditional pap and platitudes. It was a ho-hum meeting that offered nothing new until Dr. Blackburn appeared and told us about a revolutionary and exciting new concept called the Protein-Sparing Diet, or starvation with the addition of protein.

Dr. Blackburn and his colleagues at the Center for Nutritional Research at M.I.T. had done extensive work in the field of human obesity, and had observed that obese people make poor dieters but good fasters. Fasting has many advantages; primarily, of course, it accomplishes quick weight loss. However Dr. Blackburn, a short, decisive man with a staccato voice, said that even though fasters lose weight quickly, they lose too much body protein in the process. Fat can supply total body energy just as effectively as sugar, but it will neither protect nor replace body protein (as found in muscle and organs). Even the brain, which has been thought to utilize only sugar, does very well using fat. Therefore, the Center had been working on a plan for giving people between 1 and 1.5 grams of protein per kilogram of body weight daily, or just enough protein to maintain lean body mass (which is the approximate weight of your bones, muscles, organs, and essential fat). The Center recommended as the source of protein lean meat and fish. This diet *must* be supplemented with vitamins, calcium, and potassium. After the first few days of dieting, the high level of ketones acts as an appetite depressant (as supposedly also happened with Dr. Atkins' diet).

No one actually attacked Dr. Blackburn after his presentation, although you could see the disapproving glances and feel the shock at one who had dared to upset the apple cart of the traditionalists. But Dr. Blackburn remained impervious; his credentials were spotless and his data was impressive.

Several weeks later I went to Boston to meet Dr. Blackburn and to study his PSMF Diet (Protein-Sparing Modified Fast). In person, on a one-to-one basis, he was a typical, no-nonsense type of surgeon who had a remarkable understanding of the physical and psychological character of obesity. He made acceptance into the Center's diet program seem like conversion to a new religion; first you had to make a formal application for admission. (One of Dr. Blackburn's co-workers told me, "George makes it seem as if you have to walk on water to get into the program—but it's really not that hard to be admitted.") Then you were asked to give an exhaustive psychological, medical, and dietary history from the day you were born, and finally you took a psychological test, the MMPI (Minnesota Multi-Phasic Personality Inventory), to explore your anxiety levels—the Center felt (and I agree) that no one undergoing great psychological stress could be a successful dieter. If you were suffering from a major illness, divorce, or a nervous breakdown, this was not the program for you.

The food allotment—I won't call it a meal—was 9 to 12 ounces of meat or fish daily, not to be eaten all at one feeding, as well as a vitamin pill, four calcium pills, and a dose of potassium. If you felt the necessity for more solid food, you could have one-eighth of a head of lettuce, 4 radishes, half a cucumber, or pickle and clam juice. What about gnawing pains in your midsection? "Ignore them," said Dr. Blackburn helpfully, "they go away eventually."

Most of Dr. Blackburn's practice was middle-aged (I am using an obesity-based criterion of "middle age" as ranging between 25 and 50 years old), upper-middle-class, and overwhelmingly female. All had the absolute dedication he instilled in them; Dr. Blackburn was unquestionably a superstar. He preached total commitment to the cause, and would allow no failures in his program. "If you eat at parties, don't go to parties," he would say. "If you overeat on vacations, don't go on vacations," "Don't entertain, don't be preoccupied with food, and move your body "

Dr. Blackburn understood more about female metabolism than any male doctor I have ever known. He realized that overweight women simply cannot eat all they want to of anything except maybe raw vegetables, and the sooner they accept this fact, the better. Since his treatment of men and women was exactly the same (except that men got more food). I can't even call him a sexist!

I saw Dr. Blackburn make only one error while I was in Boston. He was working on an alternative diet plan called MiniMeals. These were chocolate-covered bars consisting of a balanced meal of 225 calories each. You could eat 4 bars a day and have a balanced reducing diet. Along with the MiniMeals, Dr. Blackburn conducted a total program, including weekly group therapy. But the program turned out not to be as successful as he had hoped; one day I found him somewhat disgusted. "Barbara," he said, "how *can* people eat too many MiniMeals?"

'Dr. Blackburn," I replied, "you don't understand. If you are overweight, you will eat too much of *anything* covered with chocolate, even ants!"

The Boston program was successful because success was built into the system. Motivation, modification of behavior, money, movement, and a great male authority figure all made it a tremendous success. Dr. Blackburn's cures are upwards of 60 percent, where the national average is between 5 and 10 percent. The Center did not consider you a successful dieter until you had lost your weight and kept it off for 18 weeks—"touchdown" was not counted.

At the same time as the Boston program, a Cleveland clinic was working on a protein-sparing supplement using Casec (a powdered-protein supplement) and small amounts of sugar. It used the same protein modification fasting principle, but you didn't even get to chew any meat. This diet has since become popular, but I prefer Dr. Blackburn's methods—at least he gave you something to chew on.

Dr. Blackburn wanted me to carry his teachings down to Hartford, telling me, "One good, well-motivated patient on protein sparing is better than ten patients failing on regular diets." I found it too rigorous for most of my patients; however, to this day, when I hear the Protein Sparing Fast called a "fad diet," I get furious. There has never been so much good, intelligent, practical research done on any diet. If I had any criticism of the Center for Nutritional Research, it would be that it is a very male-oriented show with, as usual, a female supporting cast. Be that as it may, I bless Dr. Blackburn for really understanding. If he had only been a woman, it would have been perfect!

11.

Dieting Is Never Fun . . .

I think the most important consideration for a doctor who is supervising dieting women is when to be flexible and when to be firm with them. I am firm when I discuss the mechanical aspects of the diet. My favorite diet, as you will see, is the Core Diet, which is a balanced-deficit diet. A balanced-deficit diet is one in which the total calories are reduced and the meals are balanced between fat, carbohydrate, and protein.

Now it is obvious that there can be all kinds of balanced-deficit diets—those with more calories than mine, for instance, and those with less. I have formulated my diet so that you get the *most* food for the most amount of weight loss. Sometimes my patients question this; but within certain ranges of caloric intake, there appears to be no perceptible difference in rate of weight loss for most overweight women. For instance, women who fast and women who eat 300 calories show the same rate of weight loss; at 500 or 700, the loss is the same; and at 700 or 900. This is true particularly at the *beginning* of a diet, when the initial reduction of calories makes the change from weight gain to weight loss

119

so pronounced. Therefore, if I can achieve maximum weight loss with 900 calories, why force my patients to eat only 700? Later in the diet I will be forced to curtail their caloric intake sharply and to nitpick about every piece of food in order to maintain the speed of weight loss—but I do this only when the patient has trouble losing weight.

Therefore, I am firm in wanting to know exactly what my patients eat, and there is no better way to know than to tell them myself exactly what they can have. However, you will note that I try to make my Core Diet simple. If I want you to eat a *piece* of fruit, I really want you to eat a common unit—an apple, a banana. I don't want you to get caught up in the mechanics of counting grapes. Cherries, blueberries, and strawberries are so seasonal that I deal with them as they appear.

I am firm about wanting patients to have a real sense of commitment about their dieting, always to be prepared to diet, and to *anticipate* what the day will bring that might interfere with their diet. Anything less than this spells failure.

I am flexible, however, in that I realize that all diets cannot be identical. A woman's body, and her life, are undergoing continual change, and her needs constantly change to meet new demands. I will go from balanced-deficit diets to protein-sparing diets to low-carbohydrate diets, depending on the situation, which is why I insist that my patients tell me about changes in their lives. Tell me when you have the flu; let me know if you can eat only liquids; call me when your ulcer acts up; you can continue to lose weight in the face of all these challenges, but you need to know how to meet the new circumstances, and you must be decisive about continuing to diet despite them.

I am also flexible about trying any new diet "breakthrough" that comes along. I think I have worked out the most efficient ways for females to lose weight today, but that doesn't mean that tomorrow something better might not appear on the scene. If it does, then I assure my patients they will have it. So far, all the stories about pills that melt fat, non-amphetemine drops that kill hunger, or some unique new combination of substances that will make weight loss effortless have proven to be false. But I'm convinced it's important to give every new theory a fair hearing, and to keep an open mind about the possibility of miracles, even if medical orthodoxy makes no allowances for them.

12.

A Calorie Is a Calorie Is a Calorie (or, Is It?)

I am somewhat disenchanted with the calorie as a reliable unit. We are taught that it is an exact measure; however, once it enters the overweight body, unusual things begin to happen. Consider the following:

> It takes 3,500 excess calories to make a pound, but–
> It takes 4,000 calories to lose a pound.

Yet in the study I discussed earlier, in which volunteers were fed 8,000 calories a day, they gained only 30 pounds when they should have gained over 240 pounds if *calories were all that mattered.*

And conversely, if calories were the whole story, why, after a meal of about 8,000 calories (the average Thanksgiving dinner), does an obese person gain 4 pounds of fat when she has consumed enough calories for only 2 pounds?

Again, why do you lose so much better eating 1,000 calories divided into three feedings than eating 1,000 calories at one feeding?

121

And when an overweight woman on a 600-calorie diet plays an hour of hard tennis, why doesn't she lose a minimum of 2 pounds per week? She should be in negative calorie balance.

Finally, why can't an obese female lose as well on 600 calories of carbohydrates as compared with 600 calories of protein? (Despite what the experts tell us, she does *not*).

There are a lot of loopholes in calorie counting and one must come to the conclusion that calories are useful only as an approximate measure when consumed by obese females. The inefficiency in the way we convert food, especially carbohydrates, to energy sets us apart from men and thin women. That is why, while I don't disregard calories entirely in my diet, by keeping the carbohydrate level low and protein level high I am able to get a steadier weight loss than if I counted calories alone. Instead of thinking of calorie requirements, think of your body requirements.

13.

Balanced Deficit Dieting

My Balanced Deficit Diet has evolved primarily from my experience in treating thousands of overweight women, but also from my own life as a dieter. Because I try to understand the total woman and tailor a diet to her needs, *which are always changing,* my dieting patients have been more successful at losing weight than those of most of my male colleagues. Bright and capable as they are, their mistake is that they fail to see that women who diet are different from men, and insist on treating women as though female physiology and psyches were identical with their own.

The decision to diet in a female, like the decision to have an orgasm, is made from the neck up. Once this decision has been made, the physician's role is to guide, understand, sympathize, and help both medically and psychologically in every way possible, not merely mechanically to hand out diet sheets and weigh the patient at fixed intervals. Diets are simple; overweight people complicate them. I tell my patients that they must follow the "triad of truth" in dieting:

1. Decrease your total calories.
2. Balance your feedings.
3. Eat (or drink) 40 to 50 percent of your calorie requirements in the form of protein (meat, fish, eggs, cheese, or liquid protein supplement).

1. Decrease your total calories.
Most obese females need to eat 1,000 calories or less a day to lose weight; it is not uncommon for women to have to cut back to 800, 600, or even 400 calories in order to lose. Many women come to me despairing because they can't lose weight on 800 calories a day—they are sure that something is wrong. I tell them that they are *normal* overweights, and that they simply have to eat less. I suggest cutting back 100 calories per day until they find the level where the weight loss is 2 pounds per week; sooner or later they will find that level. But I'm afraid the day of the balanced 1,200-calorie reducing diet is over for the chronically overweight woman, unless you balance your feedings.

2. Balance your feedings.
One of the most common diet mistakes I see is that committed by the one-meal dieter. She gives up breakfast either because she is not hungry then or because she claims that if she eats in the morning she is hungry all day long. She gives up lunch because she wants to save calories or doesn't have time to eat. She packs all her calories into one meal, in the evening. And she doesn't lose weight! Then she gets panicky and comes to my office complaining bitterly, "I'm only eating one meal a day, and I am still not losing weight!!" Scientists used to think this happened because all the calories were eaten in the evening when the body was tired and didn't burn off excess calories. My observations of women who work nights, eating at 10 p.m. and working until 8 a.m., is that they still don't use all the calories they ate in that one large meal, even though they follow it with activity. My current feeling is that the female body cannot metabolize more than a certain number of calories per meal; if the dieter exceeds this number, she will store the excess as fat. In the resting state this maximum number of calories is probably 500. The following study illustrates this point:

Given a basic 1,000-calorie diet, volunteers were divided into four groups. The first group divided the calories in the following way·

Calories
250   breakfast
250   lunch
500   supper

These people lost about two pounds per week.

Calories for the second group of volunteers were divided like this:

Calories
0   breakfast
250   lunch
500   supper

Although there were 250 fewer calories in this diet, weight loss was still about 2 pounds per week.

A third group ate the following way:

Calories
0   breakfast
0   lunch
500   supper

This group also lost about 2 pounds per week, even though they were eating 500 calories per day less than Group One.

Group Four also ate all their calories at supper:

Calories
0   breakfast
0   lunch
1,000   supper

This group lost ½ to 1 pound per week or less.

I duplicated this study in my office with similar results, except that my Group Four did not lose any weight at all. The implications of this study are clear, and amazed doctors and patients alike.

It appears that the body can efficiently handle only a *certain number* of calories per meal. If you exceed that number, as did Group Four, your body will store the excess calories as fat. My patients argue that if they give up two meals, they must be using stored fat, so that even if excess calories were stored after one big meal, they should make up for the gain by the two meals they missed. My answer is that the body has no computer which says you can eat later because you are not eating now. My patients who save all their calories for the nighttime meal *might* lose weight if that meal is 500 calories or less, but if it is over 1,000 calories they will surely store fat. If you must binge, make up for it the day after. *Don't* think you can save up for it by cutting down earlier the same day. It just doesn't work.

Group Four was frustrated and angry at losing only a small amount of weight or no weight when they were eating only one "normal" meal a day, and groups Two and Three felt like martyrs, giving up meals, cutting down at others, and still not losing as much as they thought they should. Group One was the only one that did not feel they were being cheated. We reach two conclusions from this study:

(a) Three small meals a day are better than one large meal.

(b) Saving calories for the nighttime is not an efficient way to lose weight.

3. Eat 40 to 50 percent of your caloric requirements in protein. There are many reasons for keeping your protein intake high while you are dieting:

1. Both protein and carbohydrates give off four calories of heat per gram. There is evidence, however, that the overweight female is able to handle protein food more efficiently than carbohydrate food. This means that her weight loss will be greatly facilitated if she eats more protein and less carbohydrate while dieting.

2. Protein has a high specific dynamic action (SDA). This means that a great number of calories are required to break down protein for use as energy in the body. Carbohydrates, on the other hand, use very few calories to produce energy. Since protein uses about 30 percent of its caloric value while converting to energy, a woman can eat 30 percent more protein than carbohydrates or fat, and still stay within her caloric limits.

3. Protein keeps your blood sugar at an even level, thus preventing an outpouring of insulin which might arouse hunger.

4. Protein satisfies your hunger, while carbohydrates stimulate it.

5. Protein is the *only* nutritional requirement in the body that can't be met by stored fat.

Protein (especially beef) does, however, contain many calories, because it is usually mixed with fat. An eight-ounce piece of steak may have as many as 800 to 1,000 calories. Although the female body handles protein more efficiently than any other food substance, she cannot eat all she wants even of this. An overweight female can't eat all she wants of *anything*, except perhaps raw vegetables and a few low-carbohydrate cooked vegetables.

As you now know, overweight females *cannot* handle

carbohydrates. Instead of converting them efficiently into energy, such women tend to store them as fat, and the older the female, the more pronounced this tendency becomes. I noticed this intolerance to carbohydrates when I changed my patients' breakfasts from eggs to cereal; the calorie count was the same, but the weight loss slowed considerably. I also noticed that when my patients ate extra fruit, weight loss slowed more than the additional calories could explain.

Fat, the third major class of food, packs 9 calories per gram, over twice the calories of carbohydrates and protein, and it is often hidden in food. Vegetables served in a cafeteria contain 50 calories worth of vegetable and 100 calories worth of fat. Sandwiches in the school lunchroom always have butter on them, worth an extra 100 calories. I call this the 'Hidden Hundred'—you are ingesting these extra calories without realizing it.

THE CORE DIET

I call my favorite basic diet for women the Core Diet. This is a *balanced-deficit* diet: it balances protein, carbohydrates, and fats, and creates a calorie deficit. It is simple, allows for few exchanges, and is quite strict. I call it the Core Diet because I can build on it or take away from it very easily, and it's important to keep dieting uncomplicated because the most effective diets are the simplest ones.

I always ask my patients to try to make believe this is their first diet, and to pretend they know nothing about dieting, otherwise they may become confused among the various diet programs. Each one has its own idiosyncracies, and they are not meant to be mixed and matched. I also tell my patients that the more they change the diet or invent their own variations, the less successful they will be in losing weight. If they knew how to lose weight following their own ideas, they wouldn't be in my office in the first place.

The weight lost during the first two weeks of a diet, although it doesn't represent 100 percent true fat loss, is the largest single quantity of weight lost in the diet at any one time; fortunately, because the psychological value of this phenomenon is enormous. On my Core Diet, most normal obese females lose between 7 to 10 pounds in the first 2 weeks, about 50 percent of which is extracellular water. This water is released when you cut your carbohydrate intake and use up your short-term sugar stores in the liver; some of it will come back after you stop dieting.

I always dictate my Core Diet to my patients, making them write it down themselves to avoid any possible misunderstanding and assure them I haven't left anything out. I am very emphatic that *nothing* is to be eaten that is not listed in the diet.

CORE DIET (weeks 1 and 2)

Breakfast
2 oz. orange juice or 1 orange
1 egg (prepared any way)
1 piece melba toast

To Drink Any Time
Coffee with regular milk and sugar substitute
Tea
Diet soda
Tomato juice
Skim milk

Lunch
4 oz. meat or fish
1 piece melba toast
1 cup of salad with diet dressing OR
1 cup cooked vegetable (except corn or peas)
1 fresh fruit (no grapes or cherries) OR
½ cup Jell-O (regular)

Meat: Beef (cooked weight) which can include roast beef, lean hamburger, cube or minute steak, round steak, tenderloin.
Chicken or turkey: which includes crisp skin and chicken roll.
Fish: Seafood, all canned fish, drained of oil. Fresh fish—may use 1 teaspoon of margarine in preparation.

Supper
Exactly like lunch *EXCEPT*
1 more ounce meat *AND* you can have both salad and a cooked vegetable
You can have all the salad you want at dinner, plain or with vinegar.
Diet dressing only, limited to 1 or 2 tablespoons.

Between Meals
> Raw vegetables
> Dill pickles
> Diet gelatin
> Mushrooms (raw or broiled—not cooked in butter)
> 1 can stewed tomatoes

Condiments: All—mustard, catsup, horseradish, relish, herbs, spices, soy sauce—as long as they contain no fat
Onions as a garnish

You are required to eat only three of the foods on my list: orange juice, 9 ounces of meat or fish (eggs or cheese if you are a vegetarian: half the amount of cheese—4½ oz.—and 1 egg for every 2 oz. of meat). Orange juice is a must, even though it is high in sugar, because it is a good source of potassium, and you lose a lot of body potassium the first two weeks of a diet.

I allow eggs to be cooked in 1 teaspoon of margarine because my patients, like most heavy women, do not like eggs and find them most palatable if they can have them scrambled. Women who like eggs, on the other hand, have been brainwashed by male doctors into avoiding them because of their cholesterol content. Unless you have a high cholesterol (over 300 mg percent) or a very marked family history of heart disease, I don't see why eggs should be curtailed; severe coronary artery disease is not a problem for women before menopause. (This is one area where women have a distinct advantage over men.) No cause and effect has yet been demonstrated between cholesterol intake and cholesterol blood level in the otherwise healthy female, or between cholesterol blood level and heart disease. As a good source of protein that costs only 75 calories, I like to use eggs in my diet. If you don't want eggs, however, you eat nothing else for breakfast. Initially I allow no substitutes.

One piece of melba toast per meal is a token gesture for those who want bread.

I feel that a good fluid intake is necessary, but I don't set any limit on what my patients should drink, except for cautioning them that tomato juice is the *ONLY* juice they can have in unlimited quantities. Fruit juices are nature's most concentrated form of sugar (next to whole milk). Nearly everyone believes that grapefruit juice can't be fattening because it is sour. I explain to my patients that there is plenty of sugar in grapefruit juice, and that if they want to drink something sour, they can have lemon juice.

I allow nonfat milk in unlimited quantities; most women will not drink very much of it. In the first place, it isn't as interesting when not accompanied by a piece of chocolate cake. Secondly, as we become adults many of us lose the enzyme needed to break down the sugar in milk, so that milk sugar is dumped directly into the large intestine, giving rise to gas, bloating, constipation, and/or diarrhea; many of my patients who do drink a lot of milk complain of digestive troubles.

I can see no reason for not drinking all the coffee, tea, or diet soda you want. You can't gain weight on any of them.

I specify 4 ounces of meat at lunch and 5 at dinner not necessarily to make my patients weigh their portions, but so they will realize there is a limit. Four ounces is an average restaurant serving, and is equivalent to 7 ounces of raw meat. If a patient who insists that she is following the diet is not losing, I will ask her to weigh the meat. The problem is usually the other way; my patients are more likely to eat *smaller* portions of meat or fish than I allow, particularly at noon. Patients also complain about how inconvenient it is to get plain meat or fish. Since I tell them always to be armed with a can opener and a can of chicken, tuna, or salmon, I cannot take this complaint seriously. Almost any restaurant and any greasy-spoon lunch counter will sell you a plain hamburger (without the bread) and some lettuce.

Many women ask why I use regular Jell-O as a dessert. My answer, if I am going to allow them fruit, is that a sugar is a sugar. Jell-O is a refined sugar expanded with water; fruit is a natural sugar expanded with water. The calories are quite similar, the only difference is in the fiber content of the fruit.

Substitutions
I allow only one real substitution during the first two weeks of the Core Diet. (Since I do offer choices among the required foods, I don't think this is too harsh.) I allow 1 alcoholic beverage to be substituted for 1 fruit—but it must be *nonsweet*, and not wine or beer.

It can be any whiskey mixed with tomato juice, club soda, diet soda, or water. Some people would rather have a drink than eat a piece of fruit, and the substitution does not seem to do any harm. I think people who feel pressured to drink socially at a business lunch or dinner are as silly as the people who feel they have to eat socially, but if this substitution will make them more comfortable, it does not damage their diet. Alcohol is a different kind of nutrient (that is, neither protein, fat, nor carbohydrate) called ethanol. It has 8 calories per gram, and is quickly and easily absorbed by the body and immediately burned for energy; it does not get stored as fat.

Brown-Bagging It

I am a firm believer in using canned goods when dieting, especially if you have to pack lunch or supper. All the equipment you need is a can opener, salt and pepper shakers, and a plastic fork.

You should have on hand at all times:

MEAT

Canned boneless chicken
Canned boneless turkey
Fish
Tuna (water- or oil-packed)—drain oil
Sardines (mustard and tomato)—drain oil
Mackerel
Salmon

VEGETABLES (use small cans; these are eaten cold)

Asparagus
Mushrooms
String beans
Beets
Belgian carrots
Zucchini

But NO canned fruit (even without sugar); no canned Spam; no corned beef.

Fresh fruit is very easy to carry with you, as are hard-boiled eggs and cheese.

There is no reason why it shouldn't be as easy to assemble a brown-bag lunch as it is to stand in a cafeteria line.

Second Two Weeks of Diet (weeks 3 and 4)

If the patient has lost between 7 and 10 pounds the first two weeks of the diet, I add the following to the Core Diet:

1 extra ounce of meat at lunch
2 extra ounces of meat at supper (for variety, these may now
include lamb, veal, and liver)
1 extra egg (as part of a meal or snack)
1 oz. hard cheese (as part of a meal or snack)

If she has lost less than 7 pounds, I will allow no additions, but will
adjust the existing diet. What are the reasons a patient might fail to lose
from 7 to 10 pounds the first two weeks?

1. She didn't follow the diet exactly.
2. She had been dieting previously and had already lost her
free water (which amounts for up to 50 percent of the 7- to 10-pound
loss).
3. She is taking the birth-control pill or some other medication
that favors fat formation or increases fluid retention.
4. She is a true hypometabolic and needs thyroid.
5. She has lost and gained weight so frequently in the past year
that it will take her longer to get started (see Re-Start Diet.)
6. She has chosen the highest-calorie alternatives on the
menus, such as beef, bananas, and beets, instead of chicken, celery, and
citrus fruit.

Third Two Weeks (weeks 5 and 6)
If weight loss is between 3 and 4 pounds on weeks 3 and 4, I then add
a few more foods. The purpose of this is to keep the dieter interested,
and also to see how far I can push the food load up and still maintain a
1½- to 2-pound per week weight loss.

One Chinese Meal from a Restaurant
(Yes, I said Chinese meal—but not soup and no sweet 'n' sour.)
If you do not like Chinese food, you can have instead one cup
of rice per week.
The Chinese meal follows the same idea as the "treat meal"
that used to be popular among diet doctors—that is, you diet 6 weeks
and then you can have "one free meal" as a reward. Actually, I include
it not as a treat meal, but to show my patients that certain basic
principles of eating can be learned by eating Oriental cuisine. They are:
1. no bread at meals
2. main dishes which use small amounts of very lean meat

3. bulky, low-calorie vegetables
4. no rich sauces or gravies
5. very light batter for frying
6. emphasis on texture and color of food, instead of quantity
7. minimal desserts

This is the kind of food you can eat when you stop dieting and go on maintenance. The monosodium glutamate used as seasoning may cause a two- or three-pound weight gain on the scale the day after eating Chinese food, but this is just fluid retention.

At this stage of the diet, I also add two alcoholic drinks a week, and you don't have to exchange anything for them. The ban is also lifted on wine at this point.

Fourth Two Weeks (weeks 7 and 8)
If weight loss has continued at 1½ to 2 pounds per week, then I add:

1 dipper ice milk *or* sherbet instead of fruit, twice a week *and*

cold cuts (ham, pastrami, bologna) in half the quantity of regular meat (2 oz. cold cuts to 4 oz. of regular meat) for one meal a day only

I allow sherbet or ice milk because their calorie count is not that far above fruit and gelatin, and when my patients eat out, this satisfies their need for dessert. There is a definite risk, however, that one dipper of sherbet will trigger a craving for more sugars—although I have never had a patient go on a binge on sherbet alone.

Usually I stop adding food at this point. If my patients are still losing weight at the rate of 1½ to 2 pounds per week, I might allow 1 piece of high-fiber bread at lunch or breakfast instead of melba toast. I add cottage cheese only if my patients agitate for it, and then only in exchange for both meat *and* fruits. I find weight loss slows down when I add cottage cheese to a diet. That's about as much food as I will add and still call it a *diet*.

Then I might start to go backward—yes, *backward*—if the rate of loss (1½ to 2 pounds a week) decreases. As a female loses weight, she consumes less energy when she moves, and this will be reflected in a slower weight loss unless either exercise is increased or food intake is decreased. This starts becoming very apparent after from 15 to 20 pounds have been lost. In my experience, it is easier to decrease food than to increase exercise; my patients prefer starving to moving. So I subtract all the extras I added and return to weeks 1 and 2 of the Core Diet. If weight loss still isn't satisfactory after 2 weeks, I take away the melba toast, then 1 piece of fruit, then the second piece of fruit. Meat is reduced by 1, 2, or 3 ounces per day. In the end, you might be left with this:

Breakfast
     1 egg
     coffee
Lunch
     3 oz. meat
     salad
Supper
     4 oz. meat
     salad

The only alternative to eating less is more sustained exercise.

If you break your diet at parties or on vacations, you only waste time, and it is time that defeats most diets. If you must break a diet, do it on cheese, meat, or eggs; try to avoid bread and sweets.

My Core Diet is strict, but it works. You have nothing to figure out or wonder about—I've done it all for you! I tell my patients, "I don't want you to think about food at all, because that's what got you in trouble." I aim to eliminate any need to make decisions about food. The only food a dieter requires is meat and vitamins; everything else is superfluous, and is included only to relieve the monotony. I ask my patients to listen to what I say, not to what they think their body is saying, if they want to be successful dieters. If they modify what I say or change it to suit their inclinations, they will lose weight less efficiently. I do use vitamin supplements with this diet; the more I cut down on the food, the more vitamins I tend to use. Usually one multivitamin pill with iron once a day is sufficient.

THE HIGH CALORIE WEIGHT LOSS DIET FOR YOUNGER (under 25), ACTIVE WOMEN AND PREGNANT AND LACTATING WOMEN

This group of women can handle more food—because they haven't developed the carbohydrate intolerance of older females (old in the weight-losing sense, that is), and because they require more calories for their own growth (teenagers) or for fetal growth (pregnancy). I used to put all my patients on this diet because it's so effective, but too many chronically overweight women stopped losing weight after the first few months. So now I reserve it for this group only.

THE HIGH CALORIE WEIGHT LOSS DIET

Breakfast:
>4 oz. orange juice or 1 serving of fresh fruit
>1 or 2 eggs (cooked any way) and 1 piece of dry toast *or*
>1 cup cold unsugared cereal and skim milk

Beverages
>Same as Core Diet

10:30 A.M.
>8 oz. tomato juice *or*
>8 oz. skim milk

Lunch:
>1 sandwich (using 2 pieces of bread, not a roll)
>The filling must be either beef, chicken, turkey, or tuna (1 tablespooon mayonnaise allowed to can of tuna)
>lettuce and tomato (if desired)
>catsup or mustard
>fresh fruit or Jell-O (1 serving for dessert)

3:30 P.M.
>8 oz. tomato juice *or*
>8 oz. skim milk

Supper:
>All the beef, chicken, turkey (broiled or baked), or fish you can eat at one sitting
>All the salad you want, with diet dressing or oil & vinegar
>All cooked vegetables, except corn or peas
>1 serving fresh fruit or Jell-O for dessert (may be saved for later)

Free foods are the same as in the Core Diet, except that you may add boiled cabbage.

There are certain rules that must be followed in this diet:
1. The bread cannot be eaten at the supper hour. It must be eaten by noon or you forfeit it.
2. Meat *must* be eaten at one sitting—you can't save it for later.
3. You can't save *both* fruits for the evening.
The break snacks are important here, as they help you stay full; and also keep the digestive system working at peak efficiency.

This may seem like a lot of food, but the diet works well for active women under 25 and pregnant women. It is also a great diet for men.

A natural slowing down of meat consumption usually occurs after the novelty of eating all you want wears off. It's a "Catch 22"—you can have all you want, but you don't want all you can have. Most overweight women are not particularly fond of meat anyway; the exception is the teenager, who can eat enormous quantities of it.

Weight loss tends to be somewhat slower on the High Calorie Weight Loss than the Core Diet (except for men), but since those who are on it are in periods of growth, speed of loss is not crucial. Young, active females will still lose weight very much as with the Core Diet. If weight loss doesn't take place as it should, the first food to be eliminated is the bread—that usually starts things moving again. The additions in subsequent weeks parallel those of the Core Diet.

THE RE-START DIET

One recurring problem in my practice is that of the patient who cheats and comes back to the office heavier than she was on her last visit. She may go away for a vacation and put on a pound or two because she couldn't resist sampling an exotic cuisine; or she may have found Thanksgiving and Christmas too much of a temptation. Whether it's happened once or several times, it is quite difficult to (1) get her to go back on the *same* diet; (2) get her started losing weight again.

I hate to put these patients on a different diet; I've found my Core Diet to be the best. However, something strange has unquestionably happened to their already slow metabolisms, because when they go back on the Core Diet they lose weight much more slowly than they did in the beginning. It takes about six frustrating weeks for them to pick up the weight-loss tempo that they had when they first started on the Core Diet; it's almost as if those stubborn fat cells know that if they hold out long enough, they will get fed.

I was puzzling about this when I remembered the principle of inertia: that it takes six times as much energy to restart a system that has stopped running than it does to keep it going. Since burning fat involves energy, I postulated that it must take a bigger push to get started again on a diet when you've fallen off the wagon than it does to stay on it. So I dropped my returning teenagers from the High Calorie Weight Loss Diet to the Core Diet, and that seemed to get them started again with a minimum of delay. But I dropped my adult female patients from the Core Diet back to what I now call my:

RE-START DIET (weeks 1 and 2)

Day 1:
> Nothing but liquids—coffee, tea, diet soda, tomato juice, skim milk, bouillon, diet gelatin
> Drink throughout the day, or divide into meals, if you prefer

Day 2:
> Add 9 oz. meat or fish (no pork or lamb)—broiled/baked/boiled
> Divide into at least 2 feedings *plus* the liquids

Day 3:
> Add 2 large salads with diet dressing to the food in Days 1 and 2

Day 4:
> Add breakfast of 2 oz. orange juice, 1 egg, and ½ holland rusk to all previous days

Day 5:
> Add 1 piece of fresh fruit to all previous days

Day 6:
> Add second piece of fresh fruit to all previous days

Day 7:
> Repeat Day 6 for 8 days

This is usually extremely effective, both in breaking the patient's eating-hunger cycle and in giving her a rapid weight loss; often it is even more rapid than her initial weight loss was. I sometimes repeat this diet on a two-week cycle until the patient is at the weight she had reached when she stopped her first diet. Then I proceed with the Core Diet, or the High Calorie Weight Loss Diet.

EXTREME DIETS

As I've said before, overweight people can fast more easily than they can diet; this is the principle underlying Dr. Blackburn's Protein-Sparing Modified Fast. The idea of this diet is to supply just enough protein to replace body needs, and nothing else. The diet calls for 1 to 1.5 grams of protein per kilo of body weight, or about 9 to 10 ounces of meat or fish per day for an average female. This is combined with potassium, calcium, vitamins, folic acid, water, coffee, tea, and diet soda.

The protein must be divided into two feedings. Menus might look like this:

Lunch	**Supper**
4 oz. chicken roll *or*	cube steak *or*
3½ oz. can tuna	shrimp cocktail
(water packed) *or*	(mustard sauce) *or*
4 oz. grilled ground round patty	lobster tails (2)

This is the quickest and simplest way I know to lose weight. A liquid version of the original protein-sparing diet has become popular in 1977; it allows you to drink a protein supplement instead of eating meat. It is easier, but it puts an end to meals as social occasions, and it has also become a gimmick, with excessive prices being charged for the liquid supplement. I use the liquid form of protein only when my patients insist on it, or when they have to cut back their calories to a level where they can't have 9 ounces of meat or fish. I prefer a liquid protein that has 30 calories per tablespoon and supplies 7½ grams of protein. Protein sparing is a safe and quick way to lose weight, provided you are followed closely by a physician.

CRASH DIETS

Crash diets are crazy diets that people go on just to break the monotony of sensible dieting. Usually they reduce calories too, but in a gimmicky way; these diets give you all you want of one or two foods, but you don't want to eat all you can have. Crash diets teach you nothing except that you can lose 10 pounds in two weeks eating almost any crazy combination of foods you want, as long as you reduce your total calorie count below your usual intake. Of course, one-half the weight loss is water, which is why you regain the weight immediately when you begin to eat normally. (But what is normal eating for an overweight, anyway?) Think of some of the crash diets you may have been on: skim milk and bananas, or steak and eggs and tomatoes, or cabbage soup and rice. The value of crash diets lies in their kookiness and their speed. They are of no value to the older overweight or to someone who has more than 15 pounds to lose.

My patients often ask me about this or that silly diet they read about in magazines. One day I got very disgusted with all these foolish diets and said to myself, "The next patient that comes in, I will put on the wackiest diet I can think of." My next patient was a young girl of twenty-two, moderately overweight (she needed to lose about 15 pounds). She was the perfect candidate for a ridiculous diet because her opening statement was, "Don't give me one of those boring low-calorie diets. I'm sick of them, and they are too much trouble."

"I have just the diet for you," I told her, "the Sunshine Diet!" (It was very sunny that day.)

"It sounds great!" she said. "What do I eat?"

"The beauty of it," I said, "is its simplicity. One orange and one 2-ounce hamburger for breakfast, two oranges and two 2-ounce hamburgers for lunch, and three oranges and three 3-ounce hamburgers for supper. Take 1 multivitamin pill along with this."

"Super!" she exclaimed. "It's so simple!" She left the office very excited. When she returned in two weeks she was overjoyed with her 10-pound weight loss, but I felt guilty.

"Listen," I said, "that diet was a joke. It was only meant to show you that you can lose weight on any program that reduces calories, at least for the first two weeks."

"I don't care," she said, "it was a great diet." So I started to normalize the diet, taking away one orange and adding a salad. She came back two weeks later with the expected additional weight loss of about four pounds, but she was unhappy. "I liked the Sushine Diet better," she pouted. "That was fun." I explained again that it was a hoax I had made up on the spur of the moment, but she didn't want to listen.

The strangest part was that all that week people kept calling me up asking to be put on my "Sunshine Diet." To this day, whenever I talk about it, it evokes a great deal of interest. It's easy to underestimate the importance of variety and change to keep dieters interested. Most diets are monotonously alike, and women *need* a few gimmicks to stimulate the imagination.

GOAL SETTING

There comes a time in every diet when the patient must decide how much she wants to weigh and when she is going to stop dieting. One day the patient will ask, "Doctor, what *should* I weigh?" If I tell her her ideal weight, she will invariably say, "Oh, no, that's too thin for me." I can almost see her mind working: "That doctor wants to keep me coming to see her as long as she can." So I always counter her question with, "What would you *like* to weigh?" Unfortunately, the answer she gives is rarely the ideal weight for her, but I've learned not to argue.

Many women get caught up in clothes sizes as a way of determining their goal; I prefer using scale weight. Dress size means nothing; as I've said, there are size 12 women who weigh 160 pounds. But every patient seems to think of herself in sizes instead of pounds. When I ask my patients, "How much did you weigh when you got married?" they invariably answer, "I was a size 10." A woman who is losing weight must decide which of the following standards she wants to set her goal by:

1. scale weight
2. dress size
3. measurements
4. what her friends and relatives say

If you are a 25-year-old female who is 5'6" and weighs 175 pounds, your ideal weight is 130 to 140. But at 150 you fit into a good size 12, at 140 your measurements look almost perfect, and at 130 your mother thinks you look sick! Decide at the beginning of your diet on what index you are going to base your success, and stick with your decision.

GOOD DIETS DO END—THEY DON'T JUST FADE AWAY

A diet should have an ending, just as it has a beginning. In order to be considered successful, the dieter should have reached her goal and *held it* for at least 18 weeks. Patients who walk away from the diet before it is completed, saying, "I can do the rest myself," haven't learned anything and will usually regain that weight within a few months. A successful diet should teach you three things:

1. *what it takes to make you lose weight*
2. *what it takes to make you maintain weight*
3. *what it takes to make you gain*

If you haven't learned all of these things for yourself, as they apply to your own body, your diet has not truly succeeded.

WEIGHT MAINTENANCE

Most overweight women think there is only one good way to lose weight, but that there are many ways to maintain weight. They couldn't be more wrong. There are many ways to lose weight. All sorts of diets, from 0 to 1,200 calories per day, using many different combinations of food, will allow you to lose weight, but there is only *one* way to maintain weight loss and that is: *never eat flour and refined sugar again!* When I first heard that dictum, I thought my life was ruined. No chocolate, no bread, no cake—what kind of an existence could that possibly be? Well, it could be designer dresses, it could be bathing suits, it could be more confidence. It could be health, it could be love, it could be sexuality. Life could be good and still be carbohydrateless, and maybe even be better without the pain of being overweight. If you *must* see the future in terms of food (and it is difficult to stop the habit of a lifetime), it could be cheese, a smooth wine, a perfect strawberry (is there anyone who can resist a perfect strawberry?), a marvelous steak, a superb salad, a fluffy omelet.

Most women think of weight loss as a V-shaped curve!

"The minute I eat normally, I gain weight," they tell you. By "normally," they mean adding flour and/or sugar to their diet. That's precisely what will happen for at least six months after you've lost weight; during that period you are extremely susceptible to regaining weight from eating flour or sugar.

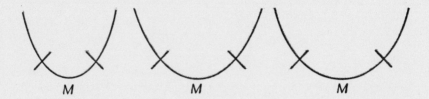

M   M   M

    Weight loss, maintenance, and gain follow a U-shaped curve; the trick is to find the middle of the U, which is weight maintenance. Unfortunately, the flat part of the U-curve can be very short in some unlucky females, and the difference in food intake between weight loss and weight gain can be amazingly small; it might be just one sweet cocktail, one piece of pie, or one helping of seafood Newburg. It's just at the moment when you think the pressure is off at last that you have to be more careful than ever.

    What makes maintenance so tough?

    1. Fat women have poor memories.
They can't remember they are fundamentally fat when they look thin. Remember, inside of every formerly fat woman—is a fat woman. Once you begin thinking that you are thin by nature, you are in deep trouble; you become overconfident and relax your guard.

    2. Weight maintenance is boring.
Losing weight is exciting. Your body changes, you buy beautiful new clothes, your friends can't believe their eyes at the transformation. Once that's been accomplished, though, it becomes a new norm, and what's normal isn't exciting anymore. People who meet you for the first time think you are naturally thin; that's no fun.

    3. You realize that people who wear small-size clothing can also have problems.
Shoulders don't fit, skirts and slacks are baggy. Your husband starts complaining about the cost of altering your old wardrobe and perhaps of buying a new one. Once you wore anything you could get into; now you are more fussy.

    4. Your friends liked it better when you were jolly and fat and always cooking goodies for them.

5. Boredom sets in, and the old habit pattern mood=food returns.

6. Your diet no longer takes precedence in your life. You are once again a mother, a wife, an employee first; the diet game is over.

7. Eating more in maintenance leads to eating more generally. A smidgeon of carbohydrates picks up the appetite and before you know it, you're off and running toward a larger size.

How do I handle weight maintenance in dieting after my patients reach their goal? First, I encourage them to lose a few pounds more than their final goal, because at the end of the diet you always owe the scale a few pounds of water weight. Remember all that weight you lost in the beginning? About one half of it was water, and you gain some of that fluid back after the diet is over.

When you reach that level of two pounds beyond goal, I have two sets of instructions. I tell the over-25 age group to eat nothing made with sugar or flour, no fruit juice, and no whole milk for the next month. I also ask them to weigh themselves daily—first thing in the morning after they have gone to the bathroom, without clothes. If there is a weight gain over two pounds, I tell them to assume that it is more than water weight and to go back to the Core Diet. They continue the Core Diet for as long as it takes to lose two pounds.

It is also wise for them to record everything they ate before they gained the weight; the results might be surprising. The oil in oil-and-vinegar salad dressing might have put them over the top, or maybe it was those six prunes for breakfast. One month of maintenance dieting will give a woman a good idea of what her body can handle. My patients then come back to the office to discuss any problems; after that I turn the monitoring of their weight over to them, cautioning them never to let it go more than 10 pounds over their goal weight.

With younger patients, my regime is not as strict. When they have achieved their goal, I allow them one serving of food containing sugar or flour per day.* It can be anything from fried chicken to pancakes, but the rest of the day they must follow the High Calorie Weight Loss Diet. They also must weigh themselves daily, following the same instructions as my older patients. If they gain over two pounds, I put them back on the High Calorie Weight Loss Diet.

I also tell all patients several things I hope that they will *always* remember: First, when you have a weight problem, you must learn to

*Besides what is already in their weight-losing diet.

deal with hunger in an appropriate manner. There are only certain foods you can eat to satisfy your hunger. If you continue to respond to hunger by eating foods you want rather than foods you can have, you will either always be hungry or always be fat. Remember: *Your excess fat cells will always be there*. They may decrease in size, but they never disappear.

Second, in order to be successfully thin, you must be vain and narcissistic. That doesn't mean you can't be kind, loving, and generous as well, but unless you develop a big ego and value the look and feel of your new body, you will lose it.

Third, if you can hold your weight down for at least two years, if and when you start to gain weight again you will gain weight like a thin person. Yes, thin people do gain weight, but only at the rate of one pound per month, or 12 pounds a year. This can easily be dealt with. A fat person gains weight at the rate of five pounds per month, or 60 pounds per year, which is exceedingly difficult, if not impossible, to control.

Overweight is a disease of *control*, not cure. This, however, does not make it any less treatable. Those physicians who turn their backs on the problem of overweight, or relegate its treatment entirely to nonprofessionals, are denying a large segment of our population much-needed help. Millions of people suffer pain, discomfort, and even despair from this disease. Even though our knowledge in this area is still limited, there is a great deal we can do nevertheless, and we owe it to our profession to take this illness as seriously as we do any other and use every available weapon in the medical arsenal to combat it.

BEHAVIOR MODIFICATION AND OTHER DIET GAMES

In the past few years behavior modification, which is a combination of Pavlovian reconditioning and common sense, has been very popular in dieting. Diet doctors have been using behavior modification for a long time, but it took the psychologists to put a label on it and make it more complex. In essence, behavior modification is an attempt to de-program the overeater: If you eat fast, learn to eat slowly. If you eat in front of the TV, eat only in the kitchen. If you eat every half hour, try eating every three and a half hours. Write down how you feel when you're hungry or full, eating or cheating. Never eat standing up. Put all food scraps in the disposal immediately.

There is no question that reconditioning is of great value in treating overweight women, but most claims of success using this method alone have been overly optimistic. However, when a patient of mine has trouble sticking to her diet, I find a few behavioral tricks can be useful. If you were my patient, I would first try appealing to the pain-pleasure principle. I would tell you first to keep remembering how painful it is to be overweight, and then:

1. Look at yourself nude in a full-length mirror from all angles. Take with you whatever forbidden food you are about to eat. If, after a long, hard look at yourself, you still feel you must eat it, eat it slowly, watching in the mirror the whole time. It usually is a terrible sight.

2. Every week, go shopping for clothes, and buy a dress one size too small. It will remind you that you are not as thin as you think you should be. But keep your old dresses, size 18 to 20—they will remind you of what you were and will be again unless you're careful.

3. Imagine your husband or boyfriend lusting after a thinner woman. If you really care more about *him* than about eating, this is the most effective ploy of all.

4. Choose friends who are skinnier than you are, but who are always dieting.

Or you can try to decrease the pleasure of eating:

1. Mash your cake, put out a cigarette in your ice cream, let the gravy and potatoes get cold. Make bread pills out of bread. If you still want these items after that, go ahead and eat them.

2. Ever see a sight that nauseated you? Associate it with food you like—melted butter on burned fingers for instance. I have a picture in my office of a luscious-looking grinder (sub, hero) with lettuce, tomato, and cold meat, and a dead pigeon lying in the middle. Nobody leaves my office and buys a grinder, at least not immediately.

3. Eat boring, dull food. The blander the better, since over-weights respond intensely to good taste. Anyone who has had to share a hypertensive's salt-free diet loses weight without trying.

4. Eat food you don't really like. Remember, you do not have to enjoy every mouthful.

5 Learn to interpret smells differently: fried chicken—greasy; Italian food—heavy and oily.

6. Learn to substitute something phony for the real thing; artificial sugar for real sugar.

7. Resolve that you'll break your diet only on gourmet food; you'll be safe 99 percent of the time.

Finally, if all these methods fail, I use my cheat-list strategy. If you must break your diet, break it my way. The rules of the cheat list are:

1. You must cheat consecutively.
2. You may skip no foods.
3. You may go backward on the list once you have eaten something on it.

For instance, to get to fruit, you must eat your way through eight categories. If you go as far as fruit on the cheat list, you will hold your own or gain only slightly. Higher than fruit and up to ice milk, you will gain about one or two pounds per week. Higher than item 16, you can gain up to five pounds per week. But remember, a cheat list is to be used only if you feel you *must* cheat; its purpose is not to encourage cheating but to give you some kind of order to your binges. It organizes your cheating. Follow it from the top every time you cheat, starting with raw vegetables. Most of my patients get to about four or five, then quit in disgust; they find it is too much trouble to work down to the food they crave. If they follow the list, there is a good chance they can successfully fight a sudden irrational desire to go off their diet.

THE CHEAT LIST

1. RAW VEGETABLES
2. COOKED VEGETABLES
3. EGGS
4. HARD CHEESE
5. SOFT CHEESE
6. FISH—CANNED, FROZEN, FRIED
7. MEAT, PLAIN
8. COLD CUTS
9. FRUIT
10. PLAIN POPCORN
11. HIGH-FIBER BREAD
12. RICE
13. POTATOES
14. FRUIT JUICE
15. CRACKERS
16. ICE MILK *OR* SHERBET

17. ICE CREAM
18. PEANUTS
19. POTATO CHIPS, ETC.
20. FOODS SERVED CREAMED OR WITH GRAVY
21. NOODLES
22. COOKIES
23. CAKE
24. PIE

Now you have them—all the diets and all the methods I use in my office to help patients. I use medicine when it is physically and psychologically necessary; there is no form of treatment I will not enlist in the interest of making my patients happier and more comfortable while they are losing weight. The willingness to change, the daring to try something new, and the courage to stand behind my own opinions, even if they are unpopular, are some of the principles I practice by. However, my foremost concern is that the healthier, happier, *and* thinner woman who leaves my office goes away with increased knowledge, more self-confidence, and a new feeling of self-worth—and also knows that the battle has just begun!

P.S. What to Do Until the Woman Doctor Comes

Chances are that when you are ready to start a diet there will only be a male doctor around to assist you. Is a male doctor better than nothing when it comes to dieting? Absolutely! You will get a physical exam and blood tests. You will get a scale and someone to weigh you; you may or may not get medication and sympathy. You *will* get a great deal of male propaganda.

Let me try to give you a clue about the care and handling of the male doctor who handles your diet. First and foremost—humor him; if you don't feed his ego, he won't diet your body. Second, keep your sense of humor. You'll need it; he may not laugh once during the long weeks of your diet. Follow his rules completely to make sure they don't work, before you try to get him to follow *my* rules. Act as if you thought up a good way to diet all on your own. If you are very clever, you can make it seem like *his* idea, and then you will have no trouble at all.

Temper his male rigidity with your new knowledge and *never* allow him to make you feel psychologically weak, inferior, or

149

ill-informed, as he will undoubtedly do at some time during the course of your diet.

Listen to his instructions when it comes to straight medical matters like a cold, a stomachache, or diarrhea. But always take his diet philosophy with a grain of salt. Let your own weight loss be the guide to your success on his diet. If you are following his diet perfectly and are losing less than one to two pounds per week (according to your size) then you must follow a different plan—Try my plan first, before you go looking for other alternatives.

But remember, if you don't like your male doctor's advice and you don't like the woman doctor's advice—don't take your own advice. That got you into trouble the first time and it will get you into trouble again. You should use the principles of female weight loss I've explained in this book to set up a diet plan of your own that will work for you. Good luck—and keep losing!

INDEX

Quiche, 44

Radishes, 117
Raisins 45
Rashes 113–14
Redistribution, 86
Relish 129
Re-Start Diet, 132, 136–38
Rice, 132 139
Rice diet program (Duke University), 70
Ricotta cheese 44
Roast beef, 128
Roller-coaster syndrome, 88–89

Salads, 128 134 135, 137
  diet dressing, 128
Salmon 130, 131
Salt, 43, 52
Sandwich 135
Sardines, 131
Sauces 133
Sauerkraut, 44, 77
Scales weight 51
Seafood 128
Sex 95–96
Sexuality 29–30
Sherbet 133
Shortness of breath, 37
Shrimp cocktail 138
Skin 84 85
Smoking, 89–90
Soda
  club 130
  diet, 44 128 130, 137, 138
Soy Sauce 129
Spaghetti sauce, 44
Spam 131
Specific dynamic action (SDA), 126
Spices 129
Spiegel Dr. Herbert, 97
Spinach 67
Steak 126 128, 139
  cube 138
Stretch marks (stria), 84–85
String beans, 131
Sugar
  refined 141 143
  substitute, 128
Sunshine Diet, 5, 139–40
''Supermothers,' 93 96
Surgeons, 107–8
Surgery 19 26
Sweeteners, artificial, 77
Sweets, 134

Tarragon 44
Tea, 128, 130, 137, 138
  camomile, 52
Teenagers, 21–25, 26, 60, 70–71, 97–99
  135, 136
Thorazine, 60
Thyroid problem, 56–59
Toast dry 135
Tomatoes, 135 139
  juice 67, 128, 129, 130, 135, 137
  sauce 44
  stewed 44 129
Toxemia of pregnancy, 79
Tranquilizers, 59, 60
Tuna, 44, 130, 131, 135, 138
Turkey, 128, 135
−20 BMR Diet, 112
Tylenol, 62

Underactive thyroid 56–57
Universal overweight personality syndrome, 29–33, 38, 96, 112

Valium 59
Veal, 44, 132
Vegetables, 128, 129, 131, 135
Vinegar 128, 135
Vitamins, 19, 117, 134, 138
  A, 68
  B, 68
  B-complex, 19
  B_{12}, 68–69
  C 19 68
  D, 68
  deficiencies, 68–69
  E 68
  K, 68

Walking, 70, 73–74
Watercress, 52
Weighing, 51
Weight
  distribution of, 85–86
  ideal for females, 10–11
  maintenance, 141
Whiskey, 130
Wine, 130
Wiring jaws, 106–7
Women and men, differences between, 2–4, 7–13

Yogurt, 45

Zucchini, 131

ABOUT THE AUTHOR

Barbara Fiedler Edelstein, M.D., was born in Massachusetts, attended Bucknell University and the Hahnemann Medical College in Philadelphia. She was a psychiatric resident (Eastern Pennsylvania Psychiatric Institute) in Philadelphia and the Institute of Living in Hartford. In 1967, as an outgrowth of her interest in psychiatry, general medicine, and overweight, she started a bariatrics practice. She has continued specializing in overweight problems ever since and has an active practice in Bloomfield, Connecticut. Dr. Edelstein lives with her husband, Stanley, a gynecologist, and their three children, in West Hartford, Connecticut.